SpringerBriefs in Agriculture

SpringerBriefs in Agriculture present concise summaries of cutting-edge research and practical applications across a wide spectrum of topics in agriculture with a fast turnaround time to publication. Featuring compact volumes of 50 to 125 pages, the series covers a range of content from professional to academic. Monographs of new material are considered for the SpringerBriefs in Agriculture series. Typical topics might include: A timely report of state-of-the art analytical techniques, a bridge between new research results, as published in journal articles, and a contextual literature review, a snapshot of a hot or emerging topic, an in-depth case study or technical example, a presentation of core concepts that students must understand in order to make independent contributions. Best practices or protocols to be followed. A series of short case studies/debates highlighting a specific angle.

Vítor João Pereira Domingues Martinho ·
Anja Vieweger · Tadeusz Pęczek · Jelena Ružić
Editors

Chemical-free Weed Control

Outcomes on Medicinal and Aromatic Plants

Springer

Editors
Vítor João Pereira Domingues Martinho
Polytechnic University of Viseu
Viseu, Portugal

Tadeusz Pęczek
Foundation for Education and Social Dialogue PRO CIVIS
Kielce, Poland

Anja Vieweger
Department of Crop Sciences
Research Institute of Organic Agriculture FiBL
Frick, Aargau, Switzerland

Jelena Ružić
CEKOM 3LJ Institution
Trilj, Croatia

ISSN 2211-808X ISSN 2211-8098 (electronic)
SpringerBriefs in Agriculture
ISBN 978-3-032-34698-8 ISBN 978-3-032-34699-5 (eBook)
https://doi.org/10.1007/978-3-032-34699-5

This Springer imprint is published by the registered company Springer Nature Switzerland AG
The registered company address is: Gewerbestrasse 11, 6330 Cham, Switzerland

Acknowledgements

This publication is based upon work from COST Action "Non-chemical weed management in medicinal and aromatic plants (MAPs) (weedingMAPs)", CA23123, supported by COST (European Cooperation in Science and Technology).

COST (European Cooperation in Science and Technology) is a funding agency for research and innovation networks. Our Actions help connect research initiatives across Europe and enable scientists to grow their ideas by sharing them with their peers. This boosts their research, career and innovation.

cost
EUROPEAN COOPERATION
IN SCIENCE & TECHNOLOGY

Funded by
the European Union

About This Book

The book presents the results of research carried out by interdisciplinary teams on non-chemical weed management in medicinal and aromatic plants, as part of the COST CA23123 project, and is of practical interest to students, academics and agro-forestry producers. More specifically, the book aims to provide further insights into the current political and legislative framework within the European Union and in each of its Member States, which may act as drivers or barriers to the implementation of non-chemical management practices and technologies in the production of aromatic and medicinal plants. The European Union has a set of strategies for climate and biodiversity, as well as a range of instruments within the Common Agricultural Policy, which are worth analysing in terms of their impact and relationship with the aromatic and medicinal plant sector. Furthermore, it is important to understand the decision-making process at farm level, and this is also addressed in the book. Indeed, there are a number of factors relating to farmers' perceptions, technological availability and needs, labour availability, consumer pressures, market functioning and the legislative framework that influence farmers' decisions. Finally, the book addresses the involvement of stakeholders in the aromatic and medicinal plant sector.

Contents

Chapter 1
Overview of the European Programming and Legislation Impacting the Adoption of Non-chemical Weed Management Strategies in Medicinal and Aromatic Plants

Artur Sobolewski, Anna Gajek, Paweł Sobczyk, and Jelena Ružić

1.1 Introduction

Weed management is generally one of the most challenging aspects of crop production. If not optimally managed, weeds pose a serious competition for nutrients, water or light and therefore often lead to lower yields and higher levels of pests and diseases. The sector of medicinal and aromatic plants (MAPs) is particularly affected by weeds, as herbal raw materials are strictly standardised according to pharmacopeia requirements, for example with regards to the content of organic contamination, as well as pesticides or other toxic residues. This creates numerous challenges for cultivation, but also further down the value chain. Although MAPs are generally grown on smaller areas than other agricultural crops, their relevance has significantly increased over the past few years. Information about the area of MAPs in Europe is very scattered. But by cross-checking information, it can be estimated that roughly 250,000 ha are currently cultivated in Europe. More than a hundred different plant species are involved. The turnover generated by these crops is difficult to estimate, as it varies greatly depending

A. Sobolewski (✉) · A. Gajek · P. Sobczyk
Foundation for Education and Social Dialogue PRO CIVIS, Kielce, Poland
e-mail: a.sobolewski@procivis.org.pl

A. Gajek
e-mail: a.gajek@procivis.org.pl

P. Sobczyk
e-mail: p.sobczyk@procivis.org.pl

J. Ružić
CEKOM 3LJ Institution, Trilj, Croatia
e-mail: ravnateljica@cekom3lj.hr

V. J. P. D. Martinho et al. (eds.), *Chemical-free Weed Control*,
SpringerBriefs in Agriculture, https://doi.org/10.1007/978-3-032-34699-5_1

on the species grown, the sector of use (herbal medicines, cosmetics, food, supplements or functional food), and the country of production. In France for example, a major producing country, the annual turnover is estimated at over 200 million Euros. It is also estimated that there are over 36,000 companies in Europe dealing with the cultivation, processing and distribution of medicinal and aromatic plants. A large part of this production takes place in disadvantaged agricultural areas like mountain areas (e.g. 20,000 ha in France's mountain areas alone), with a significant regional economic impact. Apart from the traditional market segments, the rapidly growing market of supplements and functional food in Europe poses significant opportunities for all actors in the MAP value chain i.e., farmers, local collectors, resource managers, processors and distributors. European production of MAPs is still predominantly conducted in conventional agriculture, but there has been a steady increase in the area under organic cultivation.

At the same time the use of pesticides in agriculture is increasingly criticised by society because of its impact on the environment, and this is also reflected in the purchasing behaviour of consumers. Products from organic farming have been gaining a growing market share over the past years and are now represented in almost all supermarkets.

Taking the above into account there is a strong need to establish a proper strategy for sustainable weed management for MAPs, especially with regards to the EU Green Deal, which aims to reduce chemical pesticides by 50% by 2030. This demonstrates the EU's serious efforts to tackle the problem in pesticides and bring about sustainable change in agriculture. This is complemented by further action plans for pesticide reduction in the individual EU Member States.

Farmers involved or interested in the cultivation of MAPs need clear support and easy access to training, information and exchange of experiences with regards to new technologies and the application of non-chemical tools and methodologies. However, the knowledge about benefits of non-chemical methods has not yet been widely enough spread across the different sectors and stakeholder groups. EU Member States, which are by law and by European and national policies obliged to introduce the legislative and practical measures reducing the use of herbicides, are often taking different positions, not always in clear favour of non-chemical methods.

The goal of this article is to identify and catalogue the legal and formal barriers and drivers for the adoption of strategies for non–chemical weed management (NCWM) along the MAPs value chain.

The starting point will be the European level with documents and policies such as:

- Directive 2009/128/EC of the European Parliament and of the Council of 21 October 2009 establishing a framework for Community action to achieve the sustainable use of pesticides (Directive).
- European Green Deal legislative and programme package, with special attention to Farm to Fork and Biodiversity strategies published by the European Commission on 20 May 2020.

- Common Agricultural Policy 2023–2027 (CAP 2023–2027) and its stand on issues such as the transition towards low impact farming across Europe, including organic systems.

From the European level, the research will move on to the barriers and drivers for the adoption of NCWM strategies included in Member States' implementing measures of the Directive, the Farm to Fork Strategy, the Biodiversity Strategy, and the CAP 2023–2027. In all four cases, the Member States have the obligation to implement the European legislation and the European strategic assumptions. The national cases for this research will be Poland and Croatia. The national measures are in particular:

- National Action Plans implementing the Directive (NAPs).
- The national policy positions regarding the target of a 50% reduction of the use and risk of chemical pesticides by 2030 included in the European Green Deal legislative and programming package. The reasoning behind such national positions entails barriers and drivers for the adoption of NCWM strategies at the national level.
- The Common Agricultural Policy Strategic Plans 2023–2027 (CAP Strategic Plans). The CAP Strategic Plans are the new tools combining a wide range of targeted interventions addressing the specific needs of each EU country. They are supporting the farmers in the transition towards increased sustainability in agri-food systems, in particular by introducing/entailing:
  - appropriate indicators measuring herbicide use in national farming;
  - requirements for farmers to have a plan/strategy on herbicide reduction;
  - direct payments conditioned to the use of NCWM;
  - incentives to uptake alternative techniques to herbicides.

This analysis of the European and national legislation, as well as strategical and programme measures, will give an insightful picture of current legal and formal barriers and drivers for the adoption of non-chemical approaches.

The methodology of this chapter relies first and foremost on the analysis of legal and policy documents of the EU and those of the selected EU member states.

1.2 The European Programming and the European Legislation

1.2.1 Directive 2009/128/EC of the European Parliament and of the Council of 21 October 2009 Establishing a Framework for Community Action to Achieve the Sustainable Use of Pesticides (E.U. 2009)

The Directive forms part of the EU legal framework covering pesticides and their use. According to Article 1 of the Directive the aim of the regulation is:

> to achieve a sustainable use of pesticides by reducing the risks and impacts of pesticide use on human health and the environment and promoting the use of integrated pest management and of alternative approaches or techniques such as non-chemical alternatives to pesticides.

Obviously the mentioning of non-chemical alternatives to pesticides already in the first Article, which sets the aims of the whole regulation, is a positive tendency.

The Directive does not mention explicitly the medicinal and aromatic plants, however, from the perspective of the subject in question – the following regulations are of particular importance:

(a) The definition of integrated pest management (IPM) (Article 3 number 6) understood as:

> careful consideration of all available plant protection methods and subsequent integration of appropriate measures that discourage the development of populations of harmful organisms and keep the use of plant protection products and other forms of intervention to levels that are economically and ecologically justified and reduce or minimise risks to human health and the environment

has a framing setting character and entails important general contents and elements.

The regulations stresses further that IPM shall emphasize the growth of a healthy crop with the least possible disruption to agro-ecosystems and shall encourage natural pest control mechanisms.

(b) In Article 3 ("Definitions") number 8 the Directive introduces the legal definition of "non-chemical methods", which are marked out as:

> alternative methods to chemical pesticides for plant protection and pest management (a) based on agronomic techniques or (b) physical, mechanical or biological pest control methods.

(c) Furthermore the Directive refers in the same Article 3 number 8 to the Directive's Annex III point 1., which showcases the agronomic techniques to be potentially applied. According to this Annex on „General principles of integrated pest management":

> "The prevention and/or suppression of harmful organisms should be achieved or supported among other options especially by":

- crop rotation,
- use of adequate cultivation techniques (e.g. stale seedbed technique, sowing dates and densities, under-sowing, conservation tillage, pruning and direct sowing),
- use, where appropriate, of resistant/tolerant cultivars and standard/certified seed and planting material,
- use of balanced fertilisation, liming and irrigation/drainage practices,
- preventing the spreading of harmful organisms by hygiene measures (e.g. by regular cleansing of machinery and equipment),
- *protection and enhancement of important beneficial organisms, e.g. by adequate plant protection measures or the utilisation of ecological infrastructures inside and outside production sites."*

(d) The Directive encompasses also regulations concerning the spreading of information and the promotion of non-chemical alternatives to pesticides.

According to Article 14 number 1 Member States shall take all necessary measures to promote low pesticide-input pest management, giving wherever possible priority to non-chemical methods, so that professional users of pesticides switch to practices and products with the lowest risk to human health and the environment among those available for the same pest problem. Low pesticide-input pest management includes IPM as well as organic farming according to Council Regulation (EC) No 834/2007 of 28 June 2007 on organic production and labelling of organic products (OJ L 189, 20.7.2007, p. 1.).

Article 7 of the Directive obliges additionally Member States to take measures informing general public and promoting and facilitating information and awareness-raising programmes and the availability of accurate and balanced information relating to pesticides for the general public, including also the use of non-chemical alternatives.

There are also regulations in the Directive, which do not mention neither the medicinal and aromatic plants, nor the non-chemical methods, but still have relevance to the subject in question.

(a) National Action Plans, introduced by Article 4 of the Directive, shall *inter alia* encourage the development and introduction of IPM and of alternative approaches or techniques in order to reduce dependency on the use of pesticides. The targets may cover different areas of concern, incl. Protection of the environment, residues, and use of specific techniques. The NAPs shall also include indicators to monitor the use of plant protection products containing active substances of particular concern, especially if alternatives are available.

(b) Article 5 of the Directive obliges Member States to ensure that all professional users, distributors and advisors have access to appropriate training by bodies designated by the competent authorities. The training shall be designed to ensure that such users, distributors and advisors acquire sufficient knowledge regarding the subjects listed in Annex I to the Directive entitled "Training subjects referred to in Article 5".

It is positive to notice that notions on IPM strategies and techniques, integrated crop management strategies and techniques, organic farming principles, biological pest control methods – are constituting essential elements of the Annex I catalogue (training subjects).

Concluding this overview of the selected regulations of the Directive from the perspective of the non-chemical weed management strategies in medicinal and aromatic plants, it must be stated that:

- the definition of the IPM sets the right frames and conditions for understanding and further implementing the methods for growing healthy crops with the least possible disruption to agro-ecosystems and/or methods encouraging natural pest control mechanisms;
- the introduction of the legal definition of "non–chemical methods" must be assessed very positively;
- the exemplary catalogue of the agronomic techniques to be potentially applied as non–chemical methods has the important directional character;
- the informational and promotional obligations on non–chemical alternatives to pesticides deliver a set of important assignments for the EU Member States;
- the introduction of the formal instrument of National Action Plans obliged the Member States to exercise more attention to the issues of reducing the risks and impacts of pesticide use;
- the presence of the organic farming principles and biological pest control methods in the obligatory training programmes shall contribute to the proper knowledge dissemination.

1.2.2 The Farm to Fork Strategy

The Farm to Fork Strategy is at the heart of the European Green Deal aiming to make food systems fair, healthy and environmentally-friendly. There is a need to redesign the food systems which today account for nearly one-third of global GHG emissions, consume large amounts of natural resources, result in biodiversity loss and negative health impacts (due to both under-and over-nutrition) and do not allow fair economic returns and livelihoods for all actors, in particular for primary producers. Putting the food systems on a sustainable path also brings new opportunities for operators in the food value chain. New technologies and scientific discoveries, combined with increasing public awareness and demand for sustainable food, shall benefit all stakeholders. The strategy sets out both regulatory and non-regulatory initiatives, with the common agricultural policy as one of the key tools to support a just transition.

Actually from the perspective of the non-chemical weed management strategies in medicinal and aromatic plants – three out of the four pillars of the Farm to Fork Strategy have relevance, i.e. (1) sustainable food production, (2) sustainable food processing & distribution and (3) sustainable food consumption.

The main document setting out the frameworks and providing general regulations of the Farm to Fork Strategy is the Communication from the Commission to the European Parliament, the Council, the European Economic and Social Committee and the Committee of the Regions "A Farm to Fork Strategy for a fair, healthy and environmentally-friendly food system" (Brussels, 20.5.2020 COM(2020) 381 final) (Communication from EC on the Farm to Fork Strategy) (E.U. 2020a, 2020b).

From the perspective of the subject in question the following chapters and paragraphs of the Communication from EC on the Farm to Fork Strategy have significance:

(a) Already in the introductory "Need to action" section there is a strong declaration stating the urgent need to reduce dependency on pesticides and antimicrobials, reduce excess fertilisation, increase organic farming, and reverse biodiversity loss.
(b) In the second section, i.e. "Building the food chain that works for consumers, producers, climate and the environment" the Communication from EC on the Farm to Fork Strategy emphasizes that enforcement of existing legislation, notably for pesticide use, is essential to ensure a fair transition. In the Annex to the Communication from EC on the Farm to Fork Strategy it is explicitly proposed to: (1) revise the Directive in order to significantly reduce use and risk and dependency on pesticides and to enhance Integrated Pest Management, (2) revise the pesticides statistics regulation, in order to overcome data gaps and reinforce evidence-based policy making.

But at the same time new and/or enhanced measures are being proposed by the European Commission, notably:

- stronger promotion of greater use of safe alternative ways of protecting harvests from pests and diseases;
- further encouragement of the use of alternative control techniques, such as crop rotation and mechanical weeding, which shall constitute main tools in reducing the use of, and dependency on, chemical pesticides in general, and the use of more hazardous pesticides in particular;
- enhanced facilitation of the placing on the market of pesticides containing biological active substances;
- reinforcement of the environmental risk assessment of pesticides.

The subject of non-chemical weed management strategies in medicinal and aromatic plants is indirectly referred to also in the paragraph dealing with the issue of the correlation between new threats from climate change to plant health. New innovative techniques, including biotechnology and the development of bio-based products, shall play a role in increasing sustainability and in accelerating the process of reducing dependency on pesticides.

The further important content from the section – "Building the food chain that works for consumers, producers, climate and the environment" is the organic farming, as it has a positive impact on biodiversity, creates jobs and attracts young farmers. An Action Plan on organic farming has been announced, with the aim to stimulate both supply and demand for organic products. It will ensure consumer trust and boost demand through promotion campaigns and green public procurement. This approach will help to reach the objective of at least 25 % of the EU's agricultural land under organic farming by 2030.

The Communication from EC on the Farm to Fork Strategy touches also upon the financial elements, in particular tax incentives, which are seen as the instruments potentially driving the transition to a sustainable food system and encouraging consumers to choose sustainable and healthy diets. The European Commission's proposal on VAT rates may allow Member States to make more targeted use of tax rates, for instance to support organic fruit and vegetables. As some of the vegetables had originally been used as medicinal plants and as the same compounds are present in medicinal plants and in vegetables–the boundaries between medicinal and vegetable plants are fluent. Parts of vegetables may be used to produce drugs, or may be effective via dietetic food (Fritz et al. 1993). This substantive arguments may prove successful in extending the tax incentives to medicinal aromatic plants.

The Communication from EC on the Farm to Fork Strategy points also out that practices reducing the use of pesticides through the CAP Strategic Plans will be of paramount importance. Moreover it is expected that the CAP Strategic Plans will reflect this transition away from the use of pesticides and promote access to advice. The subject of CAP Strategic Plans will be dealt with in more details in Sect. 1.2.4.

(c) Also the "Enabling the transition" section, accentuating the importance of research and innovation, as the key drivers in accelerating the transition to sustainable, healthy and inclusive food systems from primary production to consumption, relates in the indirect way to the subject of non-chemical weed management strategies in medicinal and aromatic plants.

 It is expected that Horizon Europe programs and projects, incl. a mission in the area of soil health and food will develop solutions for restoring soil health and functions. New knowledge and innovations will also scale up agro-ecological approaches in primary production and shall contribute to reducing the use of pesticides and fertilisers.

There is also the indication to introduce legislation converting the Farm Accountancy Data Network into the Farm Sustainability Data Network, with a view to collect data on the Farm to Fork's targets and other sustainability indicators, including:

- reduced dependency on pesticides and antimicrobials,
- reduction of excessed fertilisation,
- increased organic farming,
- reversed biodiversity loss.

The network will enable the benchmarking of farm performance against regional, national or sectoral averages. Through tailored advisory services, it will provide feedback and guidance to farmers and link their experience to the European Innovation Partnership and research projects. This shall improve the sustainability of operations of participating farmers, influencing positively also their incomes.

(d) The Communication from EC on the Farm to Fork Strategy concludes with the section about "Promoting the global transition". These regulations may influence the global, national and regional/local trade and commercial positions of the European entities aiming for the implementation of non-chemical weeding management approaches in medicinal and aromatic plants.

EU trade policy shall contribute to enhance cooperation with and to obtain ambitious commitments from third countries in key areas, including use of pesticides. Moreover, the EU will strive to promote international standards in the relevant international bodies and encourage the production of agri-food products complying with high safety and sustainability standards, and will support small-scale farmers in meeting these standards and in accessing markets.

The EU shall further focus its international cooperation on food research and innovation, with particular reference to:

- agro-ecology,
- IPM,
- plants' health and welfare,
- food safety standards,
- climate change adaptation and mitigation.

A more sustainable EU food system also requires increasingly sustainable practices by the EU trading partners. In order to promote a gradual move towards the use of safer plant protection products, the EU will engage actively with trading partners, especially with developing countries, to accompany the transition towards the more sustainable use of pesticides to avoid disruptions in trade and promote alternative plant protection products and methods.

Concluding, it shall be stated that:

- the aims of the Communication from EC on the Farm to Fork Strategy overlap to great extend with the objectives of non-chemical weed management strategies in medicinal and aromatic plants;
- it is encouraging to observe alternative control techniques, such as crop rotation and mechanical weeding, as well as safe alternative ways aiming at protecting harvests from pests and diseases, being pointed out as the substantive areas to be further legislated by European institutions;
- the fiscal, trade and commerce regulations, both on the individual and on the macro scale, which shall have positive implications on the economical perspective of European entities implementing the non-chemical weed management practices for medicinal and aromatic plants must be assessed very positively.

1.2.3 The Biodiversity Strategy for 2030

The EU's Biodiversity Strategy for 2030, being also a key part of the European Green Deal, is a comprehensive, ambitious and long-term plan to protect nature and reverse the degradation of ecosystems. Its ambitious targets for nature protection and restoration should lead to a better balance between nature and economic activities, contributing to a transformational change that will filter through to all parts of society, ensuring the health and prosperity of people and nature. The Biodiversity Strategy aims to put Europe's biodiversity on a path to recovery by 2030, and contains specific actions and commitments.

The main document setting out the frameworks and providing general regulations of the Biodiversity Strategy is the Communication from the Commission to the European Parliament, the Council, the European Economic and Social Committee and the Committee of the Regions "EU Biodiversity Strategy for 2030. Bringing nature back into our lives." (Brussels, 20.5.2020 COM(2020) 380 final) (Communication from EC on the Biodiversity Strategy) (E.U. 2020a, 2020b).

From the perspective of the subject of non-chemical weed management strategies in medicinal and aromatic plants the following chapters and paragraphs of the Communication from EC on the Biodiversity Strategy have significance:

(a) Point 2.2.2. "Bringing nature back to agricultural land" (part of second section "Protecting and restoring nature in the European Union", subsection "An EU Nature Restoration Plan: restoring ecosystems across land and sea") stresses the vital role farmers generally play in preserving biodiversity. They are among the first to feel the consequences when biodiversity is lost, but also among the first to reap the benefits when it is restored. Biodiversity enables the farmers to provide the communities with safe, sustainable, nutritious and affordable food, simultaneously generating the income they need to thrive and develop. At the same time however, certain agricultural practices are a key driver of biodiversity decline. This is why it is important to work with farmers to support and incentivise the transition to fully sustainable practices. Improving the condition and diversity of agroecosystems will increase the sector's resilience to climate change, environmental risks and socioeconomic shocks, while creating new jobs, for example in organic farming.

    Further on the Communication from EC on the Biodiversity Strategy emphasizes the role of agroecology as provider of healthy food, while maintaining productivity, increasing soil fertility and biodiversity, and reducing the footprint of food production. The organic farming in particular holds great potential for farmers and consumers alike. The sector creates jobs and attracts young farmers. Organic farming also provides 10–20% more jobs per hectare than conventional farms, and creates added value for agricultural products. To make the most of this potential, at least 25% of the EU's agricultural land shall be organically farmed by 2030. Actions stimulating both demand and supply of organic products, as well as incentivizing green public procurement and ensuring consumer's trust through promotion campaigns, shall contribute to reaching the above mentioned target.

    The same point 2.2.2. takes up the role the farmland birds and insects, particularly pollinators, as they play vital role for agricultural production and food security, constituting at the same time key indicators of the health of agroecosystems. The alarming decline of pollinators must be reversed. As set out in the Farm to Fork Strategy, the European Commission declares actions to reduce by 50% the overall use of–and risk from–chemical pesticides by 2030 and reduce by 50% the use of more hazardous pesticides by 2030. This must be supported by the full implementation of the EU Pollinators Initiative. A significant goal of this Initiative is to reduce the reliance on and harmful impact of pesticides. This involves: (a) sustainable use–ensuring pesticides are used in ways that are less harmful to pollinators; (b) reviewing approvals–rethinking the process for approving pesticides to better assess their risks to pollinators; (c) addressing derogations–tackling the use of emergency "derogations" and improving transparency around pesticide use in the EU.

(b) Point 3.3.1. "Business for biodiversity" (part of third section "Enabling transformative change", subsection "Building on an integrated and whole-of-society approach") offers a very interesting economical outlook.

The Commission offered its help, by utilising the existing platforms (such as "EU Business@Biodiversity Platform" and "Business for Nature"), to build a European Business for Biodiversity movement, taking inspirations from recent initiatives (such as "Business for Nature" or "One Planet Business for Biodiversity") and making this movement an integral part of the European Climate Pact. Particular attention shall be paid to measures incentivizing and eliminating barriers for the take-up of nature-based solutions, as these can lead to significant business and employment opportunities in various sectors. The Communication from EC on the Biodiversity Strategy does not specify these sectors, but it definitely shall be assumed that the sector of medicinal and aromatic plants will be positively influenced by the envisaged take-up of nature-based solutions, also in the area of weed management.

(c) All the above mentioned efforts shall be supported by the following initiatives and tools proposed by the Communication from EC on the Biodiversity Strategy:

- the EU sustainable finance taxonomy, with aim to help guiding investments towards a green recovery and the deployment of nature-based solutions–as envisaged by point 3.3.2. "Investments, pricing and taxation";
- methods, criteria and standards to measure the environmental footprint of products on the environment, including through life-cycle approaches and natural capital accounting–as envisaged by point 3.3.3. "Measuring and integrating the value of nature";
- a long-term strategic research agenda for biodiversity, as supported by the Horizon Europe programme and including a science policy mechanism for research-based options for ratcheting up the implementation of biodiversity commitments and contributions to filling knowledge gaps and finding solutions to improve the health of ecosystems and their contribution to human health – as envisaged by point 3.3.4. "Improving knowledge, education and skills".

Concluding, it shall be stated that the main aspirations of the Communication from EC on the Biodiversity Strategy, i.e. the reversion of the decline in pollinators, the reduction of the risk and use of chemical pesticides by 50%, the reduction of more hazardous pesticides by 50%, the dedication of at least 25% of agricultural land to organic farming management, and the increased uptake of agro-ecological practices–shall in direct or indirect way support the sustainable practices of the medicinal and aromatic plants cultivation.

1.2.4 Common Agricultural Policy for 2023–2027

Launched in 1962, the EU's Common Agricultural Policy (CAP) is a partnership between agriculture and society, and between Europe and its farmers. It aims to:

- support farmers and improve agricultural productivity, ensuring a stable supply of affordable food;
- safeguard European Union farmers to make a reasonable living;
- help tackle climate change and the sustainable management of natural resources;
- maintain rural areas and landscapes across the EU;
- keep the rural economy alive by promoting jobs in farming, agri-food industries and associated sectors.

It is a common policy for all EU countries. It is managed and funded at European level from the resources of the EU's budget (CAP 2025).

For the period 2023–27, the Common Agricultural Policy, which is one of the keys to securing the future of agriculture and forestry, as well as achieving the objectives of the European Green Deal, is built around ten key objectives. Focused on social, environmental and economic goals, these aims are the basis upon which EU countries designed their CAP Strategic Plans. The objectives are: (1) to ensure a fair income for farmers, (2) to increase competitiveness, (3) to improve the position of farmers in the food chain, (4) climate change action, (5) environmental care, (6) to preserve landscapes and biodiversity, (7) to support generational renewal, (8) vibrant rural areas, (9) to protect food and health quality, (10) fostering knowledge and innovation (CAP 2025).

The legal basis for the Common Agricultural Policy is established in the Treaty on the Functioning of the European Union. The CAP 2023–27 is covered by three regulations, which generally apply since 1 January 2023:

- Regulation (EU) 2021/2116, repealing Regulation (EU) 1306/2013 on the financing, management and monitoring of the Common Agricultural Policy;
- Regulation (EU) 2021/2115, establishing rules on support for national CAP Strategic Plans, and repealing Regulations (EU) 1305/2013 and 1307/2013;
- Regulation (EU) 2021/2117, amending Regulation (EU) 1308/2013 on the common organisation of the agricultural markets; Regulation (EU) No 1151/2012 on quality schemes for agricultural products; Regulation (EU) No 251/2014 on geographical indications for aromatised wine products; and Regulation (EU) No

228/2013 laying down measures for agriculture in the outermost regions of the EU.

From the perspective of non-chemical weed management strategies in medicinal and aromatic plants the most important of these legal acts is the Regulation (EU) 2021/2115 of the European Parliament and of the Council of 2 December 2021 establishing rules on support for strategic plans to be drawn up by Member States under the common agricultural policy (CAP Strategic Plans) and financed by the European Agricultural Guarantee Fund (EAGF) and by the European Agricultural Fund for Rural Development (EAFRD) and repealing Regulations (EU) No 1305/2013 and (EU) No 1307/2013 (the Regulation on CAP Strategic Plans) (CAP, 2021).

The following articles of the Regulation on CAP Strategic Plans shall be highlighted in the context of non-chemical weed management strategies in medicinal and aromatic plants:

(a) The obligation for Member States to establish and provide support for voluntary schemes for the climate, the environment and animal welfare ('eco-schemes'). Actions for a sustainable and reduced use of pesticides, in particular pesticides that present a risk for human health or environment shall constitute, where appropriate, essential elements of the eco-schemes (Article 31 point 4. f)
(b) Organic production is named as one of the 10 types of interventions in the fruit and vegetables sector to be chosen by Member States in their CAP Strategic Plans (Article 47 point 1 g).
(c) Development of organic agriculture and organic production are further named as indicators for the following EU Objectives presented in the Annex 1 to the Regulation on CAP Strategic Plans – Impact, Result, Output and Context Indicators pursuant to Article 7:

   - development of organic agriculture constitutes a Result Indicator for the EU Objective *"Contribution to halting and reversing biodiversity loss, enhancement of ecosystem services and preservation of habitats and landscapes"*;
   - organic production constitutes an Impact Indicator for the EU Objective *"Improvement of Union's agriculture response to societal demands on food and health, including high-quality, safe and nutritious food produced in a sustainable way"*.

(d) The aforementioned Annex 1 to the Regulation on CAP Strategic Plans names also further EU Cross – Cutting Objectives or EU Objectives being indirectly of relevance to subject of this Article, i.e.:

- Support of the viable farm income and resilience of the agricultural sector across the Union in order to enhance long-term food security and agricultural diversity as well as to ensure the economic sustainability of agricultural production in the Union;

- Enhancement of the market orientation and increase of the farm competitiveness both in the short and long term, including greater focus on research, technology and digitalization;

- Modernisation of agriculture and rural areas by fostering and sharing knowledge, innovation and digitalisation in agriculture and rural areas and by encouraging their uptake by farmers, through improved access to research, innovation, knowledge exchange and training;

- Attraction and sustaining of young farmers and other new farmers and facilitation of sustainable business development in rural areas.

CAP 2023–2027 has also been mobilised fully to financially support the implementation of the organic farming framework. Financial support for organics has been offered through payments for specific agricultural sectors and rural development commitments, with an additional stream of funding made available through eco-schemes and climate commitments.

In line with the CAP 2023–2027 Member States have three funding options to support organic farming (ECA 2024):

(1) Eco-schemes–payments for 1 year commitments for environment, climate and animal welfare (planned by 10 Member States).
(2) Environmental and climate commitments–payments for multi–annual commitments for agri-environmental and climate practices (planned by 24 Member States).
(3) Sectoral interventions–payments mainly for the cereals, fruit and vegetables (planned by 6 Member States).

Altogether, in the period 2023–2027, Member States have planned to allocate around EUR 14.7 billion to support organic farming in the CAP Strategic Plans.

The following two areas shall be highlighted in the conclusion:

- the incentives for Member States to undertake actions directed at organic production and simultaneously at sustainable and reduced use of pesticides; organic farming to be supported by the extended funding options;
- the horizontally approached EU Objectives connecting the sustainability of agricultural production in the Union to business factors such as farm incomes, innovativeness, competitiveness, market orientation, R&D activities and human capital.

1.3 The Cases of National Implementation of EU Regulations

After having analysed the level of European legislation in regards to areas connected in direct and indirect way to the scope of non–chemical weed management, in this chapter the selected regulations and initiatives from the two EU Member States, i.e. Poland and Croatia will be presented.

1.3.1 The Implementation of the Directive

a. The Polish National Action Plan

Polish National Action Plan (Announcement of the Minister of Agriculture and Rural Development of June 7th, 2023 on the national action plan for reducing the risks associated with the use of plant protection products for 2023–2027) (Polish NAP) makes direct reference and encourages mechanical and agrotechnical weed control for medicinal and aromatic plants (Polish NAP, p. 59). The sustainable use of plant protection products has become one of the priorities of environmental policy in Poland. Economic restructuring and modernization processes contribute to reducing pressure on the environment. Preserving beneficial organisms in field crops is one of the most important elements of biological protection, which is a priority in the search for non-chemical methods of plant protection. (Polish NAP 2023).

Further selected important and relevant characteristics of Polish NAP:

- The primary objective of Polish NAP is to disseminate the principles of integrated pest management. According to the document – the implementation of IPM principles, the main assumption of which is the rational use of plant protection products, based on the actual need to carry out the treatment, and giving priority to non-chemical methods, most effectively reduces the risks associated with the use of plant protection products (Polish NAP, p. 17).
- Plant protection programs, in order to support the implementation of IPM principles, must include not only information on chemical plant protection methods, but also guidelines for preventive actions, as well as information on non-chemical control methods (Polish NAP, p. 46).
- One of the elements to be used in IPM is the utilisation of proper crop rotation. Also important is the cultivation of resistant and tolerant varieties and the introduction into agricultural practice of alternative cultivation forms, such as: sowing mixtures of varieties and species, which allow better use of agricultural environmental resources without disturbing its biological balance (Polish NAP, p. 45).
- According to the Polish NAP – the selection of plant protection products should not only ensure the possibility of protecting individual crops, but also encourage the alternative use of plant protection products containing different active substances. In addition to chemical plant protection products, farmers should have the opportunity to use biological preparations containing microorganisms or macroorganisms or basic substances (Polish NAP, p. 58).
- Polish NAP introduces furthermore the system of collection of pesticide use data to assess risk reduction. Based on national and EU indicators, as well as other data obtained as part of the implementation of Polish NAP, an annual risk analysis shall be conducted based on trends, along with conclusions for subsequent years. The implementation of this action involves: (1) collecting and analysing data obtained during control, monitoring, and statistical surveys regarding the marketing and use of plant protection products; (2) updating and calculating risk indicators related to the use of plant protection products (Polish NAP, p. 55).
- Under action 1 of Polish NAP the system of certified trainings in plant protection products has been introduced and then further improved (Polish NAP, p. 25). Between 2017 and 2020, a total of 9967 training courses were conducted, with 233,303 participants completing them.

b. The Croatian Act on the Sustainable Use of Pesticides

The Directive has been transposed into Croatian law via the Act on the Sustainable Use of Pesticides (Official Gazette No. 46/2022) (Croatian Act) and other implementing and delegating regulations. This ensures legal enforceability and systematic implementation of sustainable pesticide practices in Croatia.

This Act regulates the National Action Plan for the Sustainable Use of Pesticides, the system of training for professional pesticide users, distributors and advisors, the use of electronic credentials of the appropriate level for access to the Phytosanitary Information System, the system of distribution and sale of pesticides, handling of pesticides, storage and handling of pesticide packaging and residues in packaging, keeping data collections and records, use and regular inspection of pesticide application equipment, informing the public and raising awareness about pesticides, measures to reduce the risk of pesticide use and the basic principles of integrated pest management, special procedures and use of pesticides, aerial application, informing the public, special environmental protection measures, reducing pesticide use and reducing risks in certain areas, protecting bees when applying pesticides, applying pesticides on public and green areas, risk indicators, providing treatment services, reporting and exchanging information, authorised professional institutions and tasks of authorised professional institutions, the Pesticide Commission, fees and violations, and supervision of the implementation of this Act and regulations adopted pursuant to this Act. The main objective of the Act is to reduce the risks and adverse effects of pesticides on human health, the environment, and biodiversity, while ensuring sustainable agricultural production.

The purpose of this Act is (1) to achieve sustainable use of pesticides, reduce risks and negative effects from the use of pesticides in a manner that ensures a high level of protection of human and animal health, environmental protection and biodiversity conservation, (2) to introduce mandatory application of the basic principles of integrated pest management for the control of harmful plant organisms and alternative approaches and techniques, such as non-chemical plant protection measures, in order to achieve sustainable and competitive agriculture. The Act serves as a roadmap for stakeholders across the agricultural sector, detailing specific actions, timelines, and responsibilities.

Key areas regulated by the Act are:

- National Action Plan (NAP), which is a strategic plan to reduce the risks and impacts of pesticides on human health and the environment.
- Mandatory training, which is regulated through a system of education and accreditation for professional users, distributors and advisors.
- Phytosanitary Information System (FIS), an electronic system of the Ministry of Agriculture through which records are kept on the sale and use of products.
- Mandatory regular technical inspections of machinery and devices for pesticide application (such as sprayers and atomizers).
- Prescription of strict special procedures and conditions for the application of pesticides to protect health and the environment.

- Prescription of mandatory and proper disposal of packaging waste to prevent soil and water pollution.
- Stipulation of mandatory and proper disposal of packaging waste in order to prevent soil and water pollution.

The implementation of the Act is defined in more detail through accompanying regulations available in the Central Catalogue of Official Documents of the Republic of Croatia, namely: the Regulation on Training for the Safe Handling and Proper Application of Pesticides, the Regulation on Additional Requirements for Pesticides Intended for Professional and Non-Professional Users, and the Regulation on Integrated Pest Management, which encourages more environmentally friendly production methods.

The law encourages the application of integrated pest management (IPM) and the adoption of non-chemical alternatives, aligning Croatian policy with the principles of sustainable agriculture in the EU. Integrated pest management (IPM) is a mandatory strategy for the control of harmful organisms in agriculture, based on a combination of environmentally and economically acceptable measures, while chemical pesticides are used only as a last resort. According to the Act on the Sustainable Use of Pesticides and the corresponding Regulation on Integrated Pest Management (OG 65/2025), all professional users in Croatia are obliged to apply these principles in everyday production.

The implementation of integrated environmental management (IPM) follows strictly in practice eight legally prescribed steps:

1. Prevention and control by using crop rotation, resistant varieties, proper fertilization and hygiene measures to prevent the occurrence of diseases and pests.
2. Monitoring harmful organisms through regular field inspections of crops, setting pheromone traps and monitoring forecasts.
3. Using economic thresholds of damage as a basis for making decisions on the justification of treatment.
4. Giving priority to non-chemical, biological, physical and mechanical protection measures whenever possible.
5. Using only specific pesticides that cause minimal damage to beneficial organisms and the environment.
6. Reducing the amount and frequency of treatments to the necessary minimum by local or partial application.
7. Changing active substances with different mechanisms of action so that pests do not develop resistance (anti-resistance strategy).
8. Checking and documenting the effectiveness of the applied measures based on records and the current situation on the plot.

Main advantages of the system:

- Sustainability–preserves the natural balance of the agro-ecological system and protects beneficial insects such as bees.
- Health–reduces pesticide residues in food and drinking water.

- Cost-effectiveness–optimizes costs by preventing unnecessary and expensive chemical treatments.

The Act in particular:

- introduces detailed restrictions on pesticide use in sensitive or protected areas, aiming to preserve ecosystems and drinking water sources;
- provides procedures for special uses of pesticides, ensuring that any exemptions follow strict risk reduction principles;
- outlines comprehensive data collection and reporting requirements, ensuring transparency and supporting national and EU monitoring obligations;
- establishes risk indicators and mechanisms for exchanging information with EU bodies, supporting harmonized pesticide risk evaluation;
- introduces rules on authorization reviews, ensuring periodic reassessment of compliance with safety and environmental standards.

Data collection, record keeping, obtaining data and information constitute a key chapter of the Act on the Sustainable Use of Pesticides, which ensures full traceability and control over the trade and application of plant protection products in Croatia. The entire system is integrated within a single digital framework. The central place for storing and processing all data is the Phytosanitary Information System (FIS), which is managed by the Ministry of Agriculture, Forestry and Fisheries. Access to the FIS is carried out exclusively electronically via the ePoljoprivreda portal with authentication via the NIAS system (the e-Citizens system credential). The Ministry connects the FIS databases free of charge with the data collections of other state administration bodies and legal and natural persons for easier control.

All professional users must accurately record each use of pesticides on their properties. Records must be kept in the prescribed electronic format, directly within the FIS system or via harmonised digital tables. Each entry must include the trade name of the imported or domestic product, the exact date and time of application, the dose applied, the crop treated and the size of the area. The data must be up-to-date and available to inspectors for inspection by the competent authorities at all times.

Obligations for distributors and points of sale–legal and natural persons who place pesticides on the market are obliged to monitor and report on the quantities in circulation. Distributors must keep sales records by entering data on each sale to professional users, linking the purchased quantities to the customer's FIS card number and are obliged to submit a report on the total annual sales of pesticides in the Republic of Croatia via an e-form in the FIS system of the Ministry of Agriculture.

The law obliges all participants in the chain to provide the requested information. Data from the collections are used to calculate national risk indicators and monitor trends in reducing the risk of pesticides. The Ministry regularly submits the collected information to the European Commission for the purpose of monitoring the implementation of European directives on sustainable agriculture.

While committed to achieving common EU goals, incl. The proper implementation of the Directive, Croatia faces several challenges, i.e.:

(1) economic viability–balancing reduced pesticide use with maintaining agricultural competitiveness;
(2) data collection accuracy–ensuring precise and timely reporting for effective monitoring and evaluation;
(3) awareness and education–ensuring widespread understanding and adoption of new practices among farmers.

Harmonizing agricultural production with strict environmental standards brings a number of operational and economic challenges for farmers, professionals and legislators.

Economic and market challenges include the following:

- Mechanization, and the use of non-chemical agents and biological preparations often have a significantly higher market price;
- The transition to integrated pest management can temporarily reduce the total amount of food produced and lower yields;
- Domestic producers find it more difficult to compete with imported products from countries with more lenient regulations on environmental protection and pesticide use;
- Regular technical inspections of machinery and mandatory training represent a constant financial cost.

Operational and climate challenges include:

- Climate change, especially sudden increases in temperature and extreme precipitation that favor the rapid spread of new, invasive pests;
- Due to the overuse of the same active substances, organisms are developing high resistance to permitted pesticides;
- The European Union is rapidly withdrawing traditional chemical substances from the market before effective substitutes are developed;
- Non-chemical and biological pesticides are significantly more sensitive to weather conditions such as solar radiation and rain washout and have a shorter shelf life

Technological and administrative challenges are significant, including:

- Digital (il)literacy is a significant challenge as the ePoljoprivreda system and digital FIS record keeping represent an obstacle for older farmers;
- There is a lack of sufficient agronomic advisors and field experts to quickly determine economic damage thresholds;
- The large number of small farms makes it difficult for phytosanitary inspectors to effectively monitor actual consumption in the field and control is weak..

Nevertheless, Croatia's dedication to the European Green Deal reflects a strategic vision for a sustainable future. Through robust legislation, technological tools, and a focus on IPM, the country is actively working towards healthier ecosystems and sustainable food production (Delić, 2025).

1.3.2 The Common Agricultural Policy Strategic Plans 2023–2027 (Generalized Approach)

First the aggregated information on the implementation of CAP Strategic Plans will be delivered, and then the Polish and Croatian executional initiatives will be presented.

As presented in the Sect 1.2.4 above, from 1st January 2023, support for farmers and rural stakeholders across the 28 Member States is based on the new Common Agricultural Policy legal framework and the choices detailed in the CAP Strategic Plans. The, by the European Commission approved CAP Strategic Plans, are designed to make a significant contribution to the ambitions of the Farm to Fork Strategy and the Biodiversity Strategy presented respectively in Sects. 2.2 and 2.3. They are drafted within a robust common EU legal framework and include tailored strategies to address Member States' specific circumstances and needs, since there are no *'one-size-fits-all'* solutions. The CAP Strategic Plans deploy different tools that address the Common Agricultural Policy's economic, environmental and social objectives in a balanced way. They are funded on the basis of the Multiannual Financial Framework for years 2021–2027, which allocated EUR 378.5 billion of EU-funds to CAP Strategic Plans spending programmes.

The following generalized regulations and initiatives shall be highlighted in the context of non-chemical weed management strategies in medicinal and aromatic plants:

(a) First of all it must be noted that the subject of non-chemical methods is present in the CAP Strategic Plans. To reduce the use and risk of pesticides, more than 26 % of EU agricultural land shall receive support thanks to:
   - using non–chemical methods for pest control such as precision farming;
   - banning use of pesticides in certain specific areas;
   - adopting IPM.

(b) Further on–farmers shall protect soils and preserve soil potential thanks to crop rotation. All CAP Strategic Plans include this as a new basic condition for farmers and the rotation will take place on around 85% of the arable land supported by the financial components of the CAP Strategic Plans. Crop rotations will also help disrupt pest and disease cycles and thus reduce use of pesticides.

(c) Then there is the share of the EU's utilized agricultural areas under mandatory and additional actions to be introduced by Member States and related to climate and environment.

As shown by the graph below 27% of these utilized agricultural areas shall be subjected to actions involving sustainable and reduced use of pesticides, while 10% to actions developing organic agriculture. (EC 2022) (see Fig. 1.1).

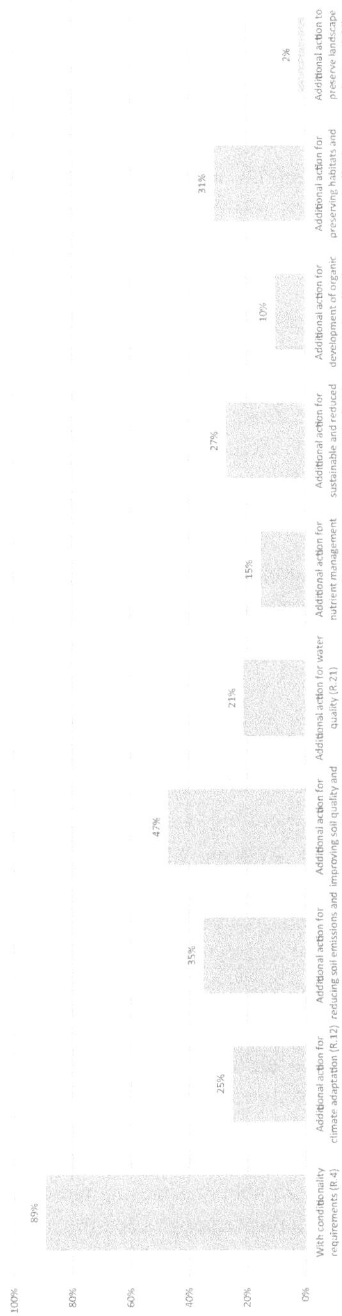

Fig. 1.1 Share of the EU's utilised agricultural area (UAA) under mandatory and additional actions related to climate and environment

(d) Reduction of pesticide and fertiliser use will also be achieved also through increasing the area of organic production. The size of the area that receives specific CAP 2023–2027 support for organic production in 2027 will almost double, reaching close to 10%, compared to the area funded in 2020 (5.6%). This will significantly help Member States' reach their national ambitions for increase in organic areas. The ambitions for those areas across Member States range from 5 to 30% in 2030.
(e) Finally there are also the economic and social regulations of CAP Strategic Plans of indirect importance for non–chemical weed management strategies:

- based on Member States' policy choices for financing across the instruments and funds (European Agricultural Guarantee Fund, European Agricultural Fund for Rural Development), 24% of direct payments are dedicated to eco-schemes and 48% of rural development spending from EU funds is earmarked for environment and climate;
- there is the support of investments in digital technologies and services to optimise resource efficiency, including for precision farming; focus is given to development of digital skills of farmers and digital tools for knowledge exchange amongst farmers, advisers, research and also to increased data sharing;
- there is also the support for advice, demonstration and training to help farmers with the transition to a more resource efficient, resilient and sustainable agricultural system.

1.3.3 The Polish Common Agricultural Strategic Plan 2023–2027

The Polish Common Agricultural Strategic Plan 2023–2027 (introduced by the Statute dated February 8th, 2023) (Polish CAP Strategic Plan 2023–2027) foresees the following measures of direct importance for the subject of non-chemical weed management strategies in medicinal and aromatic plants (Polish CAP 2023).

- Eco-schemes featuring the payments for mechanical weed control in medicinal and aromatic plants fields. The eco-schemes shall generally incentivise farmers to implement environmental and climate-friendly practices that go beyond the legal requirements and allow shift towards more sustainable and environmentally friendly agriculture. Eco-schemes are nevertheless voluntary for farmers in Poland with an allocation of 25% of the direct payments envelope. There is also specific focus on soil protection and sustainable production methods.
- Through the implementation of the Organic Farming intervention, it shall be possible to reduce the overall use of industrial means of production (including pesticides), which in turn should reduce the pressure of the agriculture sector on the environment, including on biodiversity and ecosystem services.

- Preventing damage caused by pests and weeds shall be primarily based on protection through:

- their natural enemies,
- selection of species, varieties, and heterogeneous material,
- crop rotation,
- cultivation techniques such as biofumigation, mechanical and physical methods,
- thermal processes such as solarization,
- and in the case of protected crops – shallow soil steaming (to a maximum depth of 10 cm).
- Preventing damage caused by pests and weeds shall be primarily based on protection through:

- Obligations to maintain plant health through preventive measures, in particular the selection of appropriate species, varieties or heterogeneous material resistant to pests and diseases, appropriate crop rotation, mechanical and physical methods and the protection of natural enemies of pests.
- The aim of the agricultural practice shall be to improve soil quality and the need to restore organic matter by enriching the crop structure with plant species that affect both the positive balance of organic matter and the increase of biodiversity. Furthermore, to protect biodiversity, the requirement of having at least three different crops on the farm will help limit monocultures on arable land.

The Polish CAP Strategic Plan 2023–2027 offers also free advisory and training workshops on non-chemical weed control for smallholders.

1.3.4 The Croatian Common Agricultural Strategic Plan 2023-2027

Croatia submitted its first draft of The Common Agricultural Strategic Plan of the Republic of Croatia 2023-2027 on 31 December 2021, (Croatian CAP 2023). The Commission accepted the proposal on 28 October 2022.

The Croatian strategy aims to support the sustainable development of agricultural holdings and the processing sector and to improve living and working conditions in rural areas. A number of incentives will support the protection of natural resources, biodiversity and the climate. Through investments in agricultural holdings and business development projects, it will also focus on young and small farmers and on higher animal welfare standards in the livestock sector.

The Croatian CAP (2023) reinforces efforts to reduce the income gap between small and medium-sized farms and larger ones. In order to stabilise the incomes of small and medium-sized farms, 20% of direct payments will be redistributed to all small and medium-sized farmers for their first 30 ha of agricultural land.

Support will be provided to the livestock and fruit and vegetable sectors, thereby helping farmers and rural areas in general, as maintaining production, and thus jobs, is also important for the further development of rural areas and stopping their depopulation.

Furthermore, the plan prioritises the development of and investment in quality food production in order to increase the market orientation and competitiveness of agricultural holdings. The plan supports stronger cooperation between farmers and the processing sector and short supply chains, especially for small farms and high-quality products.

The Croatian CAP (2023) aims to address a number of environmental and climate challenges facing the agricultural sector. In order to protect good soil and climate conditions, Croatian farmers must comply with new mandatory rules for the protection of organic matter and soil carbon. The plan foresees the introduction of eco-schemes, i.e. incentives for farmers who voluntarily implement practices that go beyond the mandatory environmental and climate requirements. Supported eco-schemes will ensure better crop rotation and the use of nitrogen-fixing crops (such as legumes), the maintenance of landscape features, the use of natural fertilizers on arable land, grazing on pastures and the application of methods that reduce soil erosion. In border zones along watercourses, the use of all types of fertilizers and plant protection products is limited, thereby reducing nutrient runoff. Farmers have an obligation to manage agricultural soils in a way that protects them from erosion, for example by sowing cover crops to enrich and protect the soil. In order to preserve and restore biodiversity, around 10% of agricultural land needs to be cultivated with methods that support the preservation and restoration of biodiversity. Organic farming also plays an important role in the plan. The share of organic production in Croatia currently stands at 6.5%. With the help of the CAP, the area under organic farming will increase to 12% of agricultural land.

The competitiveness of rural areas in Croatia depends, among other things, on improving the state of technical infrastructure, improving the availability of basic services and developing entrepreneurship. This will create thousands of new jobs in rural areas, along with investments in infrastructure (water, local roads and kindergartens). Support will also be granted for investments not related to agricultural holdings. Around EUR 103 million is planned for investment in rural areas under the European LEADER programme, which will contribute to social innovation.

In order to encourage young people to engage in agricultural activities, a large number of Croatian farmers under the age of 40 will receive special incentives for setting up agricultural holdings and covering initial investment costs.

Payments for improving animal welfare–pigs, cattle, sheep, goats, laying hens, broilers and turkeys raised for meat production–cover 65% of the total livestock. The measures include better nutrition and care, increasing the available space for each animal and the possibility of keeping animals for a certain amount of time outdoors.

The implementation of the plan is based on the principle of equal opportunities and non-discrimination, including access to opportunities for people with disabilities and equal opportunities for women and men.

The Croatian CAP (2023) also aims to support training and advisory activities for farmers. A register of public and private advisors has been published on a dedicated platform, allowing Croatian farmers to choose the advisory support that best suits their individual needs.

Overall, the emphasis is on digitalization and the introduction of innovations as tools to facilitate the green transition. It is planned that more than 2% of agricultural holdings will receive financial assistance for restructuring and modernizing their infrastructure, and almost half of them will receive support for the introduction of digital agricultural technologies.

Croatian CAP (2023) emphasises interventions aligned with environmental, climate and biodiversity objectives. These objectives encourage the promotion of sustainable production systems, which inherently include reducing dependence on chemical inputs, including herbicides. Although non-chemical weed control is not explicitly detailed, the environmental objectives require farmers to adopt agronomic practices – such as crop diversification, soil-friendly farming and landscape elements – that reduce weed pressure without chemicals.

The document also integrates improved conditionality, aligning with EU standards for good agricultural and environmental conditions. Although specific weed control rules are not listed in the available extracts, good agricultural and environmental frameworks–including the maintenance of permanent grassland, protection of ground cover, erosion prevention and crop rotation–directly reduce weed growth without herbicides. The document's resources highlight conditionality as a key element of environmental interventions.

The Croatian CAP (2023) also invests in interventions that preserve or enhance biodiversity–such as the maintenance of hedgerows, field borders and ecological landscape features–identified as key ecological priorities in the document. These features support natural competitors and improve habitat structure, which indirectly suppress weeds by promoting more resilient agro-ecosystems and reducing open niches exploited by weeds.

Finally, it should be noted that climate interventions represent the largest block of funding in the Croatian CAP (2023). Climate-friendly practices–cover crops, soil moisture maintenance, reduction of bare soil periods–are proven non-chemical weed control strategies. By prioritizing climate resilience, the document indirectly encourages multipurpose practices that also control weeds.

Concluding the chapter on the cases of national implementation of EU regulations the following elements shall be highlighted:

- the fact that the Polish NAP makes direct reference and encourages mechanical and agrotechnical weed control for medicinal and aromatic plants and generally gives priority to non-chemical methods, both in terms of substance and in terms of promotion, must be assessed very positively;
- the founding principles of the Croatian IPM scheme (constituting element of the Croatian NAP), i.e. healthy crop growth, justified intervention, natural pest

control and minimizing disruption–are in line with the main assumptions and aims of weedingMAPs;

- the importance of the presence in the CAP Strategic Plans of the subjects of non-chemical methods, the reduction of the use and risk of pesticides, organic agriculture–and the connection of all three subjects to the financial schemes;
- another very constructive factor are the regulations of the Polish CAP Strategic Plan 2023 – 2027 with explicit eco-schemes featuring the payments for mechanical weed control in medicinal and aromatic plants fields; this is well supplemented *inter alia* by free advisory and training workshops on non-chemical weed control for smallholders;
- the Croatian CAP Strategic Plan 2023–2027 prioritizes environmental, climate, and biodiversity objectives, encouraging farmers to adopt sustainable agronomic practices that naturally suppress weeds without relying on chemical herbicides; it strengthens EU-aligned conditionality and introduces eco-schemes that promote soil health, biodiversity, and resilient agroecosystems, all of which indirectly support non-chemical weed management.

1.4 Conclusion–Catalogue of Legal and Formal Drivers and Barriers for the Adoption of NCWM Strategies along the MAPs Value Chain

After analysing the substantive legislative, strategical and programing measures functioning at the European and national levels – the following catalogue of legal and formal drivers and barriers directly or indirectly impacting the adoption of non – chemical weed management strategies along the medicinal and aromatic plants sector's value chain is being presented as a form of a summary to the article.

– Drivers:

  1. The ambitious aims of the European legislation, incl. The need to reduce dependency on pesticides, the promotion of alternative approaches or techniques to pesticides such as non-chemical alternatives, the increase in organic farming, and the reverse of biodiversity loss.
  2. The presence in the EU legal system of the formal definition of "non-chemical methods", as alternatives methods to chemical pesticides.
  3. The presence of broad catalogue of exemplary agronomic techniques to be utilised in order to prevent or suppress harmful organisms, along with the encouragements to use these techniques.
  4. Introduction of "integrated pest management" with the objective to implement measures discouraging the development of populations of harmful organisms, to encourage natural pest control mechanisms and to minimise risks to human health and the environment.

5. The introduction of the formal instrument of National Action Plans obligating Member States to exercise more attention to the issues of reducing the risks and impacts of pesticide use.
6. Selected regulations of the National Action Plans like the presence and encouragements for mechanical and agrotechnical weed control for medicinal and aromatic plants (Polish NAP), the integration of the principles of "healthy crop growth", "natural pest control" and "minimizing disruption" (Croatian Act).
7. The presence of the subject of non-chemical methods in the Common Agricultural Policy 2023 – 2027 Strategic Plans. To reduce the use and risk of pesticides, more than 26 % of EU agricultural land shall receive support thanks to using non–chemical methods for pest control such as precision farming and banning use of pesticides in certain specific areas.
8. Introduction of actions for a sustainable and reduced use of pesticides, in particular pesticides that present a risk for human health or environment, as essential elements (however facultative) of the eco-schemes under the Common Agricultural Policy 2023–2027 Strategic Plans. 24% of direct payments are dedicated to eco-schemes and 48% of rural development spending from EU funds is earmarked for environment and climate.
9. Strong mobilisation of the Common Agricultural Policy 2023–2027 to support the implementation of the organic farming framework. The area of organic production will be increased. Financial support for organics has been offered through payments for specific agricultural sectors and rural development commitments.
10. Presence of investments' support in digital technologies and services to optimise resource efficiency, including for precision farming. The result being the extended digital skills of farmers and digital tools for data sharing and knowledge exchange.
11. Strong financial support for young farmers and generational renewal, in order to help young and new farmers with income support, the costs of setting-up their business or investments.
12. The EU sustainable finance taxonomy, with aim to help guiding investments towards a green recovery and the deployment of nature-based solutions. The presence of the financial elements, in particular tax incentives (VAT rates), to support organic fruit and vegetables.
13. The financial allocations in the EU research and innovations programs, in particular Horizon Europe, with the aim to develop solutions reducing the use of pesticides and to implement biodiversity commitments and contributions to filling knowledge gaps and finding solutions to improve the health of ecosystems and their influence on human health.
14. Trade and commerce regulations, both on the individual and on the macro scale, which shall have positive implications on the economical perspective of European entities implementing the non-chemical weed management practices for medicinal and aromatic plants.

15. Introduction of catalogue of EU objectives including (a) substantive indicators–such as development of organic agriculture, organic production (b) more economically oriented indicators–such as the viable farm income and resilience of the agricultural sector, the market orientation and increase of the farm competitiveness.
16. Measures aiming at creating data networks collecting sustainability indicators, including reduced dependency on pesticides, increased organic farming and reversed biodiversity loss.
17. The reinforcement of the environmental risk assessment of pesticides.
18. Extended catalogues of informational and promotional activities having as a subject the non-chemical alternatives to pesticides.
19. Existing obligations on EU Member States to ensure that all professional users, distributors and advisors have access to appropriate training in matters referring to the sustainable use of pesticides, incl. Organic farming principles and biological pest control methods.

- Barriers

  1. Implementation challenges of the European legislation setting the ambitious pesticide use and risk targets, as Member States prioritize economic concerns over environmental protection.
  2. The reluctance by Member States to engage meaningfully with integrated pest management implementation. Inadequate enforcement and verification mechanisms, even where formal requirements exist.
  3. Fragmented and insufficient financial Common Agricultural Policy support for pesticide alternatives.
  4. Fundamental flaws in the EU pesticide risk assessment system. The system's failures extend to industry influence in regulatory processes. Academic institutions involved in developing guidance are simultaneously working with industry, what creates a pattern of regulatory capture that compromises the independence of scientific assessment.

References

CAP (2021). The Regulation (EU) 2021/2115 of the European Parliament and of the Council of 2 December 2021 establishing rules on support for strategic plans to be drawn up by Member States under the common agricultural policy (CAP Strategic Plans) and financed by the European Agricultural Guarantee Fund (EAGF) and by the European Agricultural Fund for Rural Development (EAFRD) and repealing Regulations (EU) No 1305/2013 and (EU) No 1307/2013. Accessed 25 Sep 2025. https://eur-lex.europa.eu/legal-content/EN/TXT/?uri=CELEX%3A32021R2115&qid=1773651714766.

CAP (2025). CAP at a glance–Agriculture and rural development–European Commission. Accessed 25 Sep 2025. https://agriculture.ec.europa.eu/common-agricultural-policy/cap-overview/cap-glance_en

Croatian CAP (2023) Croatian common agricultural strategic plan 2023-2027. Off J Repub Pol

Delić I. (2025). Croatian national policy within the framework of the European green Deal and sustainable use of pesticides. The presentation delivered at the CEKOM 3LJ international conference “what about EU agricultural policies in European countries?”, September 18th–19th, 2025 in Trilj–Croatia. The conference organised within the scope of the COST action “non-chemical weed management in medicinal and aromatic plants (MAPs)” (weedingMAPs) CA23123

EC (2022). European Commission. Common agricultural policy for 2023–2027. 28 CAP strategic plans at a glance

ECA (2024). European court of auditors, special report. Organic farming in the EU–gaps and inconsistencies hamper the success of the policy. Accessed 25 Sep 2025. https://www.eca.europa.eu/en/publications?ref=SR-2024-19.

E.U. (2009). Directive 2009/128/EC of the European Parliament and of the council of 21 October 2009 establishing a framework for community action to achieve the sustainable use of pesticides. Accessed 22 Sep 2025. https://eur-lex.europa.eu/legal-content/EN/TXT/?uri=CELEX%3A02009L0128-20190726&qid=1773650423661.

E.U. (2020a). Farm to fork strategy: for a fair, healthy and environmentally-friendly food system. Accessed 22 Sep 2025. https://eur-lex.europa.eu/legal-content/EN/TXT/?uri=CELEX:52020DC0381.

E.U. (2020b). EU biodiversity strategy for 2030. Bringing nature back into our lives. Accessed 23 Sep 2026. https://eur-lex.europa.eu/legal-content/EN/TXT/?uri=CELEX%3A52020DC0380&qid=1773651267150.

Fritz D, Jakobey H, Habegger R (1993) Vegetables to be used as medicinal plants? Acta Hortic 332:105–114. Accessed 23 Sep 2025. https://doi.org/10.17660/ActaHortic.1993.332.14

Polish CAP (2023) The polish common agricultural strategic plan 2023–2027, introduced by the statute dated February 8th, 2023. Off J Repub Pol 2024:1741

Polish NAP (2023) Announcement of the minister of agriculture and rural development of June 7th, 2023 on the national action plan for reducing the risks associated with the use of plant protection products for 2023–2027. Off J Repub Pol 2023:768

Chapter 2
Gaps and Trends on Weed Management and Non-Chemical Approaches: The Specific Case of Best Practices, Drivers and Barriers in Medicinal and Aromatic Plants

Vítor João Pereira Domingues Martinho

2.1 Introduction

Weed control needs more contributions from the scientific community (Belfry and Sikkema 2018), considering its impact on agricultural performance (Kumar et al. 2024) and its relevance to improving food security (Rodenburg et al. 2014). The challenge of weed management through more sustainable practices is a dilemma for farmers, between the concerns with environmental preservation and the profitability of the farms. Often, the research is not conclusive about the positive net profitability of farming from the use of more sustainable approaches (Alm et al. 2000), which justifies the consideration financial incentives as drivers to promote greater sustainability in agroforestry systems. New technologies bring new methods to deal with infestations through optical and electrical approaches, for example, nonetheless these treatments also have several constraints related to costs, required skills, and ecological impacts (Li et al. 2026). In any case, weed management is among the concerns of stakeholders associated with organic farming and agroecology (Migliorini and Wezel 2017). These frameworks show that it is urgent to identify sustainable ways to deal with weed control, where the precision technologies (Monteiro and Santos 2022) and integrated management (Roberts et al. 2024) may bring relevant contributions, because, in some cases, the herbicides are no longer effective (Nawaz et al. 2019) and the labour use is not a solution due to its scarcity (Shekhawat et al. 2017).

V. J. P. D. Martinho (✉)
School of Agriculture (ESAV), Polytechnic University of Viseu (UPV), Viseu, Portugal
e-mail: vdmartinho@esav.ipv.pt

Centre for Environmental and Marine Studies (CESAM), University of Aveiro, Aveiro, Portugal

V. J. P. D. Martinho et al. (eds.), *Chemical-free Weed Control*,
SpringerBriefs in Agriculture, https://doi.org/10.1007/978-3-032-34699-5_2

Several non-chemical weed management strategies identified in the scientific literature (Alluri and Saha 2024) have been identified as sustainable solutions (Ofosu et al. 2023), including for biomass sources (Gazoulis et al. 2021). However, several questions remain unanswered in these contexts, which require further insights from the literature, considering the specificities of each production (Pannacci et al. 2017). In any event, it is important to combine biology knowledge, cultural practices and direct weed management (Bond and Grundy 2001). It is also relevant to take advantage of plant genetics and breeding knowledge (Debaeke et al. 2024) and to combine different methods (Hatcher and Melander 2003). The use of cover crops may be an interesting solution for non-chemical weed management, despite the potential negative effects on the main crops (Mennan et al. 2020), as well as the weed detection techniques (Wei et al. 2010). Specifically for medicinal and aromatic plants, manual control can be an option (Carrubba et al. 2009).

Considering the context described previously, this chapter aims to bring more insights into the weed management, namely through non-chemical approaches (specifically in the medicinal and aromatic plants sector), considering bibliometric analysis and information collected from Portuguese farmers. For the bibliometric analysis, several documents were considered from the Scopus database (Scopus 2026), identified in a search conducted on 28/03/2026, which were then assessed following the Bibliometrix software (Aria and Cuccurullo 2017) procedures. The bibliometric analysis was first carried out for weed management and then for non-chemical weed management.

2.2 Bibliometric Analysis

In this section, first, 10,535 documents will be analysed for the topic "weed management", and after that 77 studies for the topic "non-chemical weed management".

2.2.1 Weed Management

From a total of 10,535 documents over the period 1978–2026, Fig. 2.1 highlights the existence of 23,493 authors, 1821 sources, 504 single-authored documents, 19.2% international co-authorship, an annual growth rate of 11.83% and 4.4 co-authors per document.

The annual scientific production presents an increasing trend, namely in the most recent years (Table 2.1), highlighting the motivation of the scientific community for these topics worldwide.

The weed management topic is in the rapid growth phase (progress to saturation of 40.3%), indicating it is not yet a mature topic (Fig. 2.2), instead, it is an emergent/in development theme in the scientific literature.

Fig. 2.1 Main information

Weed Technology, Weed Science, Weed Research, Agronomy, Crop Protection, Indian Journal of Agronomy and Pest Management Science are the most relevant sources for the weed management topic (Fig. 2.3).

The most relevant authors are those presented in Fig. 2.4, namely: Chauhan, BS; Norsworthy, JK; Kumar, S; Sikkema, PH; Kumar, V; Kumar, A; Blackshaw, RE; Gerhards, R; and Kato-Noguchi, H.

Tamil Nadu Agricultural University, University of Florida, North Carolina State University, The University of Queensland, Michigan State University, Cornell University, Montana State University and University of Western Australia are the relevant affiliations (Fig. 2.5).

The USA, India, Australia, Brazil, China, Canada, Pakistan, Italy, Spain, France, Germany, UK, Iran, Japan, Malaysia, Argentina, Greece, Turkey, New Zealand, Denmark, South Africa and Israel are the most productive countries (Table 2.2).

Among the most cited countries appear the following (Fig. 2.6): USA; Australia; Canada; India; China; Spain; Italy; United Kingdom; France; and Germany. The European Union appears here represented by Spain, Italy, France and Germany.

Weed management, weed control, weed species, management practices, integrated weed, grain yield, weed density, management strategies, cover crop, and hand weeding are the most frequent words (Fig. 2.7).

The emerging topics related to weed management are the following (Table 2.3): crop productivity; soil health; food security; weed detection; deep learning; precision agriculture; detection accuracy; pesticide reduction; and severe drought.

Figure 2.8 reveals that management practices, weed density, and grain yield are niche themes (well developed, but of reduced relevance). On the other hand, weed biomass is in the middle of the relevance degree (but not yet well developed) and weed management/control is in the middle of the development degree (with a high relevance degree).

Management strategies, herbicide resistance, weed management, weed species, herbicide control, weed control, cover crops, weed suppression and weed biomass are central themes for the weed management topic (Fig. 2.9). In turn, weed density, hand weeding and grain yield are specific themes that appear in a cluster distant from the centre of the map.

Table 2.1 Annual scientific production

Year	Articles
1978	1
1979	0
1980	0
1981	1
1982	3
1983	6
1984	3
1985	6
1986	7
1987	6
1988	6
1989	8
1990	10
1991	11
1992	13
1993	24
1994	31
1995	35
1996	112
1997	94
1998	97
1999	163
2000	126
2001	130
2002	98
2003	137
2004	176
2005	195
2006	213
2007	237
2008	282
2009	233
2010	234
2011	292
2012	289
2013	324
2014	324
2015	363

(continued)

Table 2.1 (continued)

Year	Articles
2016	363
2017	374
2018	431
2019	472
2020	551
2021	622
2022	660
2023	654
2024	847
2025	1057

Fig. 2.2 Life cycle of scientific production

Figure 2.10 shows the international collaboration network, where the United States, India and Australia are the leading countries.

2.2.2 *Non-Chemical Weed Management*

Over the period 2001–2026, 77 documents were considered from Scopus for the topic of non-chemical weed management, with 300 authors, an international co-authorship

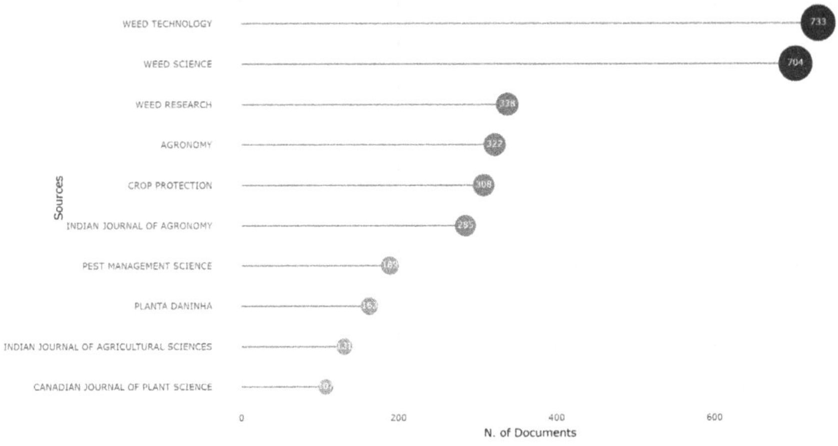

Fig. 2.3 Most relevant sources

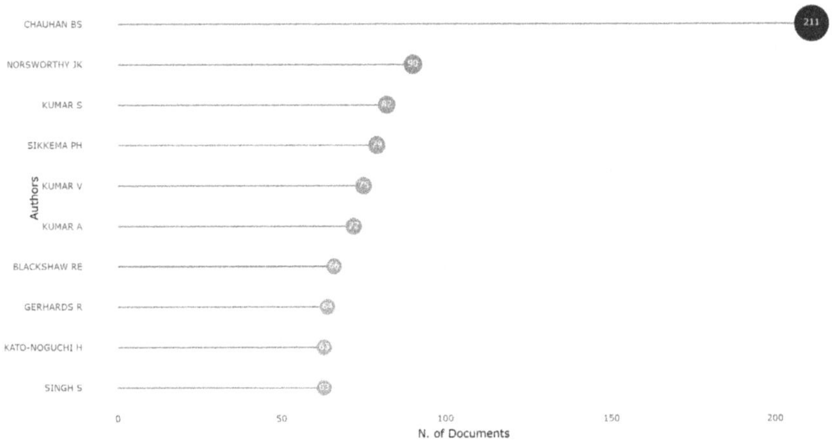

Fig. 2.4 Most relevant authors

rate of 18.18%, an annual growth rate of 4.49%, and 4.29 co-authors per document (Fig. 2.11).

The annual scientific production presents an irregular evolution over the period considered, nonetheless with an increasing trend, presenting, for example, 12 documents in 2024 and 6 in 2025 (Table 2.4).

Figure 2.12 reveals that this is a mature topic, with a progress to saturation of 73.80%, corresponding to a maturity phase. The prediction is that this topic will achieve a total of 104 documents in the saturation phase.

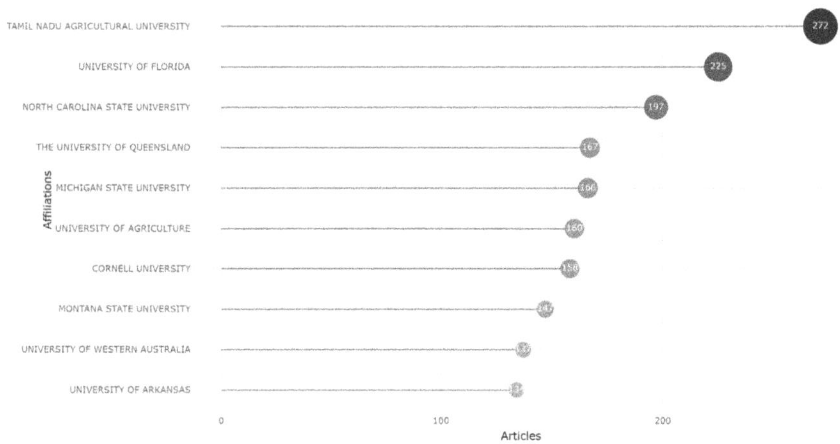

Fig. 2.5 Most relevant affiliations

Weed Research, Crop Protection, Agronomy, Horticulturae, Soil and Tillage Research, Applied Ecology and Environmental Research, Frontiers in Agronomy, and Indian Journal of Agronomy are the most relevant sources (Fig. 2.13).

The most relevant authors for this topic are the following: Barberi, P; Amini, R; Blackshaw, RE; Bond, W; Brodie, G; Chauhan, BS; Colbach, N; Fontanelli, M; Ginanni, M; and Gupta, D. (Fig. 2.14).

Tamil Nadu Agricultural University, The University of Western Australia, University of Pisa, University of Melbourne, Agricultural Institute of Slovenia, Volcani Institute, and Agriculture and Agri-Food Canada are the most important affiliations (Fig. 2.15).

India, USA, Australia, Italy, Iran, France, Israel, China, Denmark, Pakistan, Turkey, United Kingdom, Canada, Germany, Hungary, Thailand, Greece, Japan, Netherlands and Poland are the top countries in terms of scientific production (Table 2.5).

The United Kingdom, Italy, Hungary, Australia, Bangladesh, Canada, the Netherlands, China, Sri Lanka, and the USA are the most cited countries on the topic of non-chemical weed management (Fig. 2.16).

The most frequent words are the following (Fig. 2.17): weed management; weed control; non-chemical weed; hand weeding; weed density; management strategies; weed suppression; and control methods.

Cover crop, total weed, crop mixture, weed suppression, weed species, seed bank, management treatments, and hand weeding are emerging topics (Fig. 2.18).

Flame weeder, flame weeding, plant damage, weed seed, cropping systems, and weed population are emerging/declining themes (Fig. 2.19). Weed management, weed control, and non-chemical weed control are important themes, but not yet well developed. Hand weeding, weed density, and grain yield are important and well-developed themes.

Table 2.2 Top 50 countries'scientific production

Country	Documents
USA	6871
India	3870
Australia	1906
Brazil	1420
China	1263
Canada	1013
Pakistan	900
Italy	709
Spain	633
France	597
Germany	550
UK	482
Iran	444
Japan	389
Malaysia	334
Argentina	309
Greece	302
Turkey	268
New Zealand	216
Denmark	209
South Africa	189
Israel	185
Nigeria	185
Netherlands	169
Ethiopia	167
Bangladesh	165
Philippines	165
Egypt	159
Saudi Arabia	152
Switzerland	139
Indonesia	128
South Korea	119
Mexico	118
Belgium	102
Thailand	98
Czech Republic	95
Serbia	95
Hungary	94

(continued)

Table 2.2 (continued)

Country	Documents
Poland	92
Portugal	87
Sweden	82
Morocco	76
Norway	74
Kenya	62
Zimbabwe	62
Iraq	59
Tunisia	54
Austria	50
Ghana	47
Colombia	46

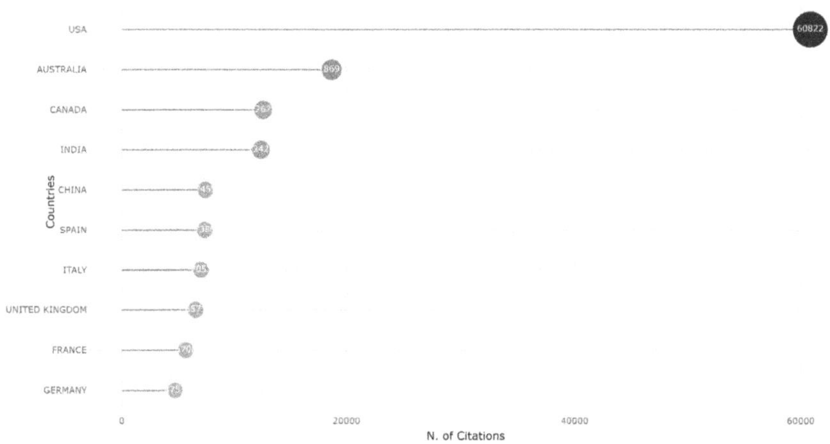

Fig. 2.6 Most cited countries

Integrated weed, non-chemical weed and weed control are central themes in this topic, appearing in a central cluster jointly with cover crops, organic farming, weed species, and herbicide resistance. Grain yield, control efficiency, hand weeding, and lowest weed are themes that appear in a specialised cluster (Fig. 2.20).

India, the USA, and Australia are key countries in the international collaboration network for non-chemical weed management (Fig. 2.21).

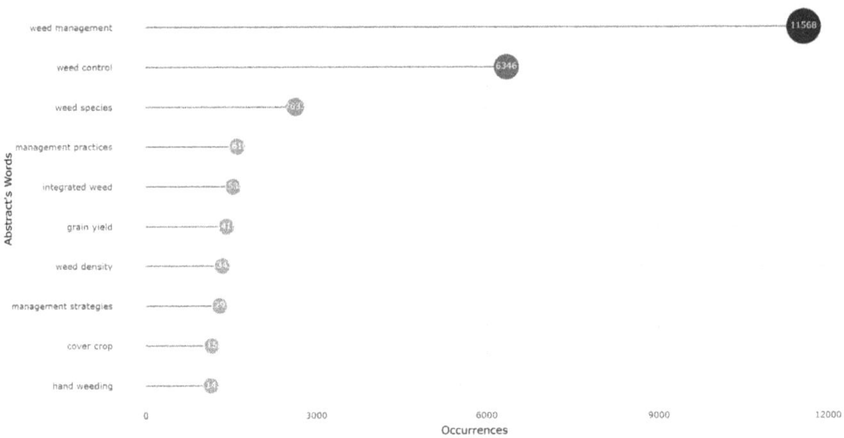

Fig. 2.7 Most frequent words (in the abstracts)

Table 2.3 Trend topics (in the abstracts)

Term	Frequency	Year (Median)
weed management	11,568	2019
Weed species	2633	2019
Management practices	1610	2019
Management strategies	1297	2020
Herbicide resistance	1004	2020
Crop production	629	2020
Block design	570	2021
Climate change	362	2021
Crop yields	325	2021
Effective weed	637	2022
Sustainable weed	457	2022
Crop productivity	236	2023
Soil health	214	2023
Food security	213	2023
Weed detection	620	2024
Deep learning	369	2024
Precision agriculture	256	2024
Detection accuracy	72	2025
Results highlight	65	2025
Significant challenge	65	2025
Pesticide reduction	15	2026
Reduction strategies	12	2026
Severe drought	11	2026

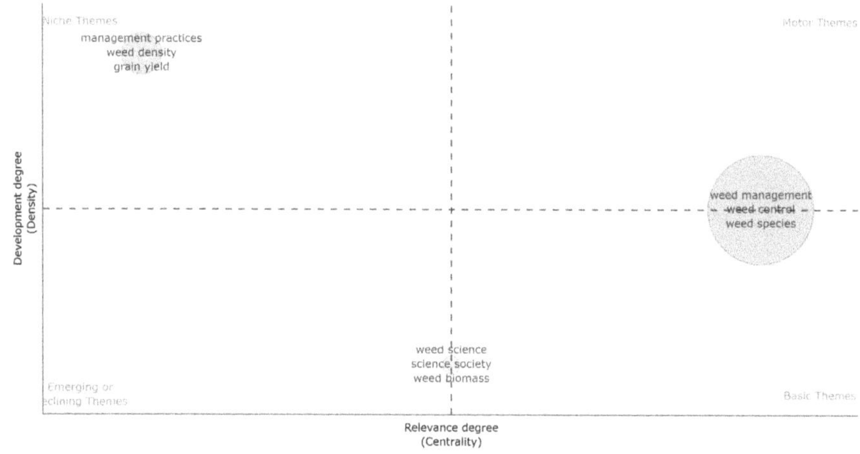

Fig. 2.8 Thematic map (in the abstracts)

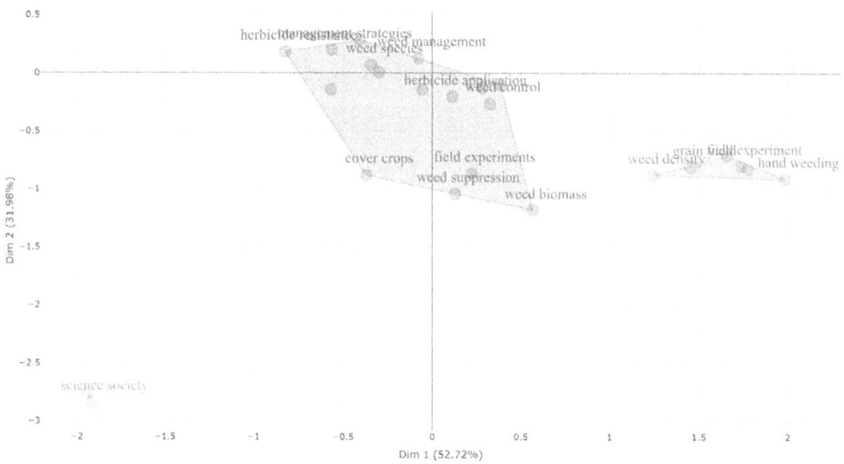

Fig. 2.9 Factorial analysis (in abstracts from multiple correspondence analysis)

2.3 Best Practices, Drivers and Barriers on Non-Chemical Weed Management in Medicinal and Aromatic Plants in the Portuguese Context

In total, 14 Portuguese MAPs operators, with activity related to medicinal and aromatic plants, were contacted to collect their opinions on best practices, drivers, and barriers associated with non-chemical weed management. These operators are distributed from the north to the south of Portugal, the majority have 1–5 ha (with the exception of one operator with 0.07 ha, another with 16 ha and another with

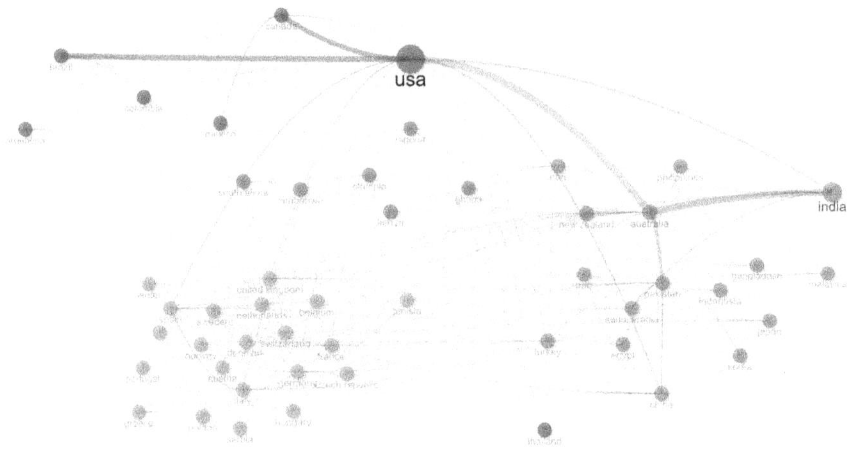

Fig. 2.10 Collaboration network

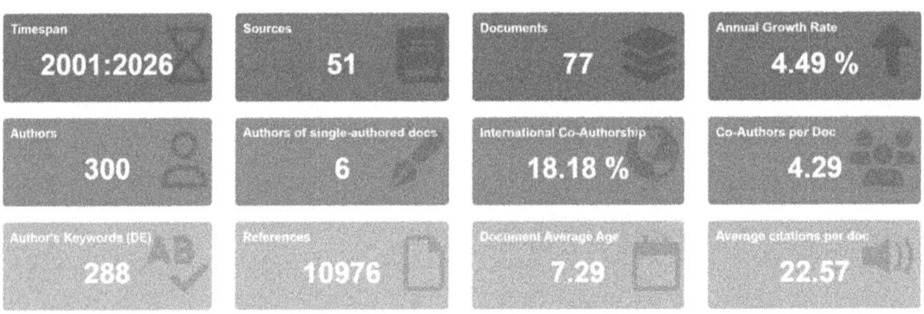

Fig. 2.11 Main information

40 ha). In general, these operators produce dried and fresh plants for export and the national market, through organic farming practices, conduct distillation of essential oils, and produce outdoors and in greenhouses. The main species produced are, for example, the following: Thymus citriodorus; Thymus vulgaris; Thymus mastichina; Mentha spicata; Mentha piperita; Corylus avellana; Rosmarinus officinalis; Aloysia citrodora; Foeniculum vulgare; Cymbopogon citratus; Cistus ladanifer; Lavandula stoechas; Lavandula officinalis; Helichrysum stoechas; Eucalyptus globulus; Ruta graveolens; Melissa officinalis; and Stevia rebaudiana.

Most of the operators agree that the environmental and health dimensions related to the use of herbicides are a concern for them. On the other hand, the majority of MAP operators have no opinion, or disagree, that optimism about the possibility of the appearance of new chemical herbicides in the future limits the adoption of non-chemical strategies. Almost all operators agree that the non-chemical weed management practices are seen as labour consuming, expensive and requiring specific knowledge. These operators suggest the following solutions to deal with these constraints:

Table 2.4 Annual scientific production

Year	Articles
2001	1
2002	1
2003	1
2004	0
2005	0
2006	0
2007	7
2008	0
2009	1
2010	2
2011	0
2012	0
2013	2
2014	1
2015	0
2016	1
2017	6
2018	3
2019	6
2020	4
2021	9
2022	7
2023	4
2024	12
2025	6

development of new technologies; more adjusted incentives from the national and European Union agricultural policies; more training courses and practical demonstration; more adjusted legislation, namely in terms of taxes; financial supports to deal with the costs of some production factors, namely the energy; protection from specific markets where the rules are not so restrictive. Half of the operators agree that public awareness of chemical residues in food influences the adoption of non-chemical techniques. The same applies to whether the social reputation plays an important role in decisions about more sustainable practices.

These operators identified the following main drivers and barriers to adopting non-chemical weed management strategies in MAPs:

- Drivers: government incentives and subsidies for sustainable practices; regulatory pressure and restrictions on herbicide use; consumer pressure for production with less use of chemicals; and additional agronomic benefits (improved soil structure).

Fig. 2.12 Life cycle of scientific production

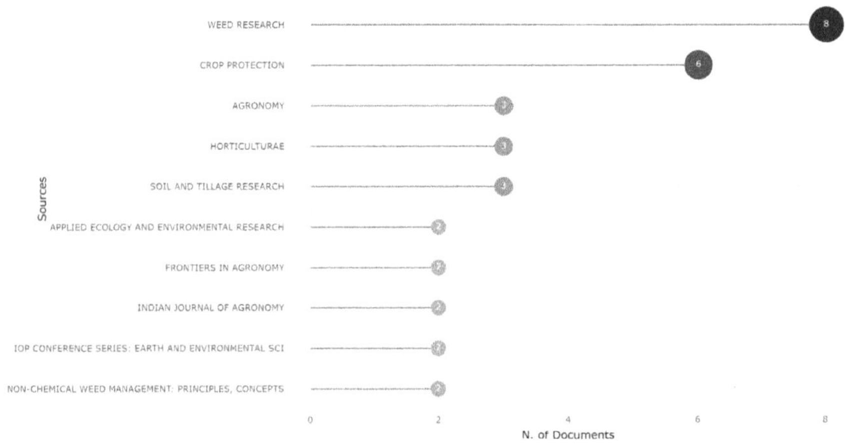

Fig. 2.13 Most relevant sources

- Barriers: high initial cost of mechanical equipment; lack of technical knowledge; greater need for labour; difficulty in large-scale application.

In general, these operators are involved in the following activities to develop and implement non-chemical strategies:

Fig. 2.14 Most relevant authors

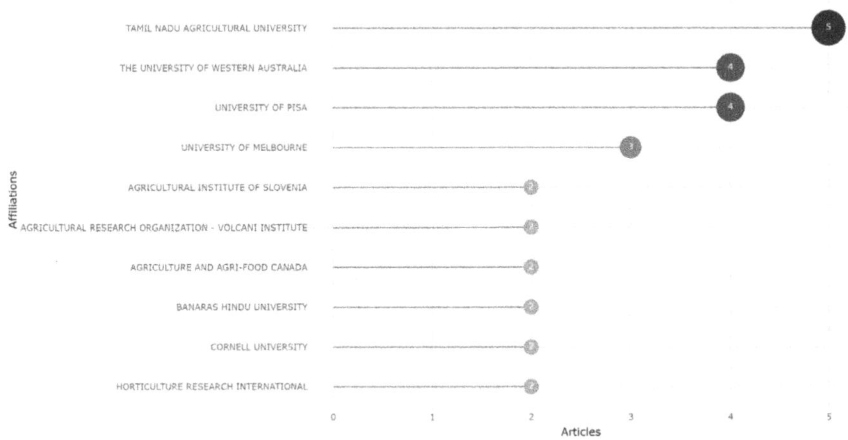

Fig. 2.15 Most relevant affiliations

- In development: participation in field trials with researchers; co-design of techniques and equipment adaptations; contribution of practical/local knowledge on management; and prototype testing (smart weeders, mechanical machines).
- In implementation: continuous monitoring of weeds; adaptation of techniques to the specific reality of the farm; training and management of labour for non-chemical practices; and planning and implementation of crop rotations.

The most effective approaches to knowledge transfer about the economic benefits of non-chemical techniques are the following:

Table 2.5 Top countries'scientific production

Country	Documents
India	31
USA	24
Australia	22
Italy	12
Iran	11
France	10
Israel	7
China	5
Denmark	5
Pakistan	5
Turkey	5
UK	5
Canada	4
Germany	4
Hungary	4
Thailand	4
Greece	3
Japan	3
Netherlands	3
Poland	3
Slovenia	3
Spain	3
Lithuania	2
Switzerland	2
Bangladesh	1
Brazil	1
Czech Republic	1
Egypt	1
Indonesia	1
Ireland	1
New Zealand	1
Saudi Arabia	1
Serbia	1
Slovakia	1
Sri Lanka	1

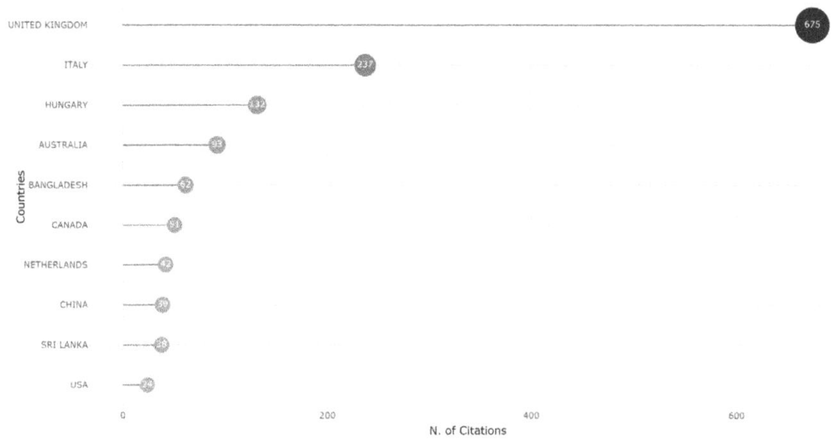

Fig. 2.16 Most cited countries

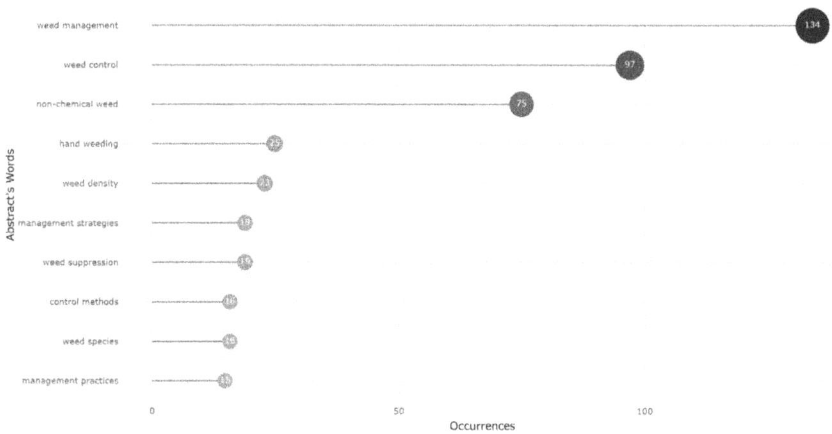

Fig. 2.17 Most frequent words (in the abstracts)

– Demonstration field days; participatory trials on farms; practical workshops; and machinery demonstrations with economic and financial feasibility analysis.

The environment for innovative start-up companies can be created through partnerships with farmers, testing infrastructure, targeted funding/financing, and links with research.

The opinions about the roles of associations and demonstration farms on the processes of communication and dissemination are the following:

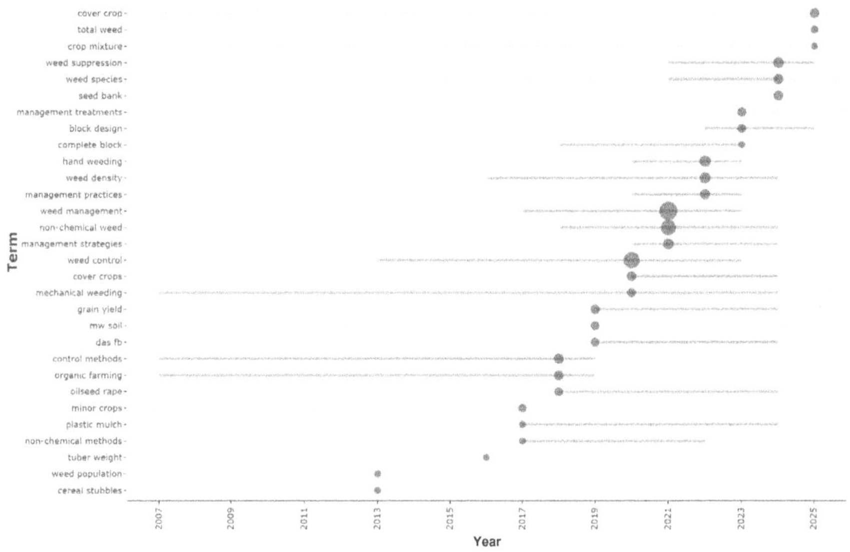

Fig. 2.18 Trend topics (in the abstracts)

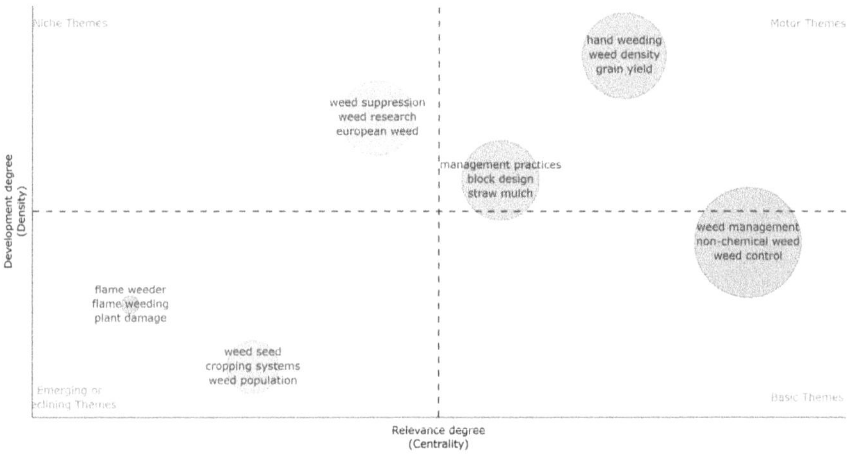

Fig. 2.19 Thematic map (in the abstracts)

- Farmers' associations: facilitate communication between farmers and researchers; offer workshops, courses, and seminars; organise meetings, forums, and conferences to exchange experiences; and create learning networks and regional/local working groups.
- Demonstration farms: demonstrate techniques (crop rotation and plant cover) in practice; conduct comparative trials between traditional and innovative methods;

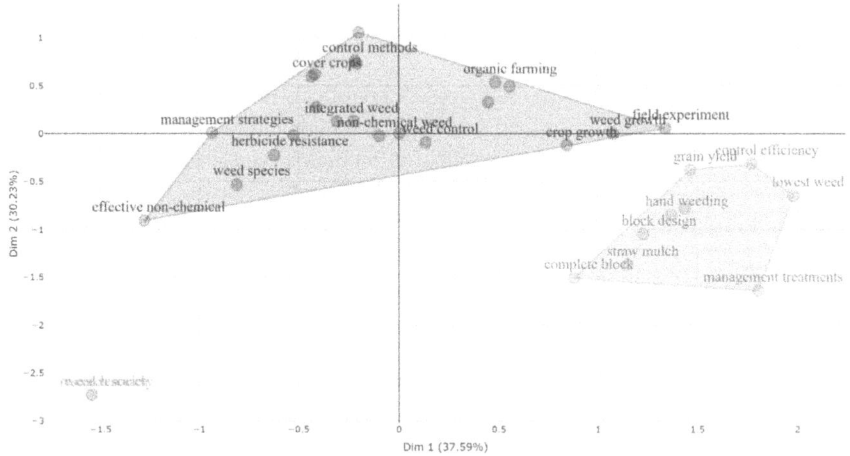

Fig. 2.20 Factorial analysis (in abstracts from multiple correspondence analysis)

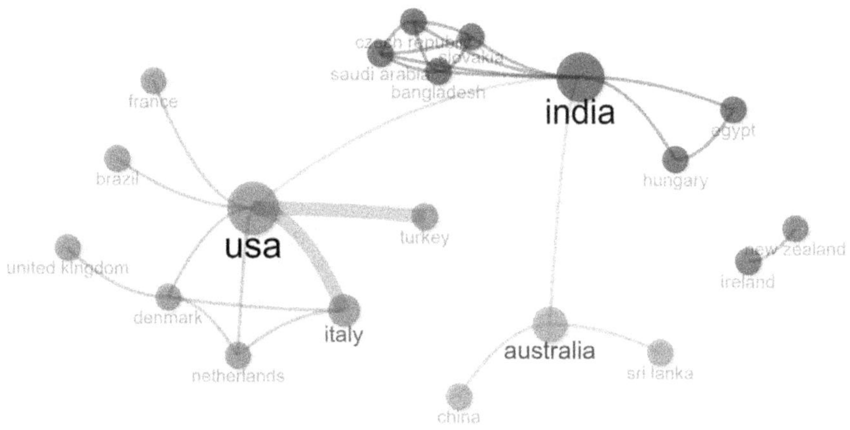

Fig. 2.21 Collaboration network

host technical visits and field days for practical experiences; and validate technologies and best practices (show productivity and cost results).

The engagement of young farmers at schools and universities can be made through curriculum integration, experimental and demonstration fields, workshops, practical activities, and integration into research projects.

2.4 Discussion and Conclusions

The scientific community has contributed significantly to understanding weed management, but there is still a field to be explored in the literature, namely about the specificities of non-chemical weed control in medicinal and aromatic plants. Even more, when the weed management with sustainable practices is not an easy task because of impacts on farms' profitability, the availability of adjusted technologies, the skills required to adopt innovative approaches, and labour shortages. The policy framework may contribute significantly to these contexts, namely by providing incentives to overcome the financial constraints inherent to the adoption of non-chemical techniques. In any case, the current context calls for greater sustainability in weed management, due to herbicide resistance and its impacts on the environment and human health. Some of the sustainable solutions are related to integrated approaches to manage weed evolution, a combination of knowledge from different fields, digital innovations for prediction, and cover crop techniques.

The bibliometric analysis highlights that weed management is still an emerging topic, and the related documents have been published mainly in specialist journals (Weed Technology and Weed Science). The United States, India, Australia, Brazil and China are the top 5 most productive countries in the world on this topic, but the most cited have been the United States, Australia and Canada. Integrated weed, grain yield, and weed density are among the most frequent words, showing the importance given to these themes in the weed management topic. On the other hand, crop productivity, food security, weed detection and pesticide reduction are emerging topics, revealing that new technologies may make a relevant contribution, and that there is concern in the documents related to weed management regarding sustainability. Specifically, about the non-chemical weed management topic, the bibliometric assessment reveals some maturity in the evolution of the associated literature. The sources where the related documents appear are also mainly from the speciality, or related fields (Weed Research and Crop Protection), and India appears in this case as the most productive country, but the most cited is the United Kingdom. Cover crop, weed suppression, and hand weeding are emerging topics, and grain yield is an important and well-developed theme.

The information collected from the Portuguese MAPs operators shows that they are concerned with sustainability, but they are also worried about the costs associated with non-chemical approaches, the related technologies, policies, legislation, taxes, and exposure to international markets with different rules. In turn, they identified government incentives, legislation to reduce herbicide use, consumer pressure, and agronomic benefits as drivers of the adoption of non-chemical techniques. The constraints are associated with costs, a lack of skills, labour needs, and difficulties in applying at large scales. The farmers' associations and the universities may play an important role by promoting courses, events, meetings, project involvement, and demonstration techniques.

The research carried out shows that promoting more sustainable weed management practices is a concern for the scientific community, nonetheless more contributions are needed to understand better the frameworks associated with non-chemical weed management in MAPs. In any case, in this study relevant findings, suggestions for best practices, drivers and barriers were identified, which may be of interest to policymakers, farmers and future research.

Acknowledgement This publication is based upon work from COST Action "Non-chemical weed management in medicinal and aromatic plants (MAPs) (weedingMAPs)", CA23123, supported by COST (European Cooperation in Science and Technology). This work was funded by national funds through FCT—Fundação para a Ciência e a Tecnologia I.P., under the project CESAM-Centro de Estudos do Ambiente e do Mar, references UID/50017/2025 (doi.org/10.54499/UID/50017/2025) and LA/P/0094/2020 (doi.org/10.54499/LA/P/0094/2020).

References

Alluri G, Saha D (2024) Emerging Perspectives on Non-Chemical Weed Management Tactics in Container Ornamental Production in the United States. Hortic 10. https://doi.org/10.3390/horticulturae10030281

Alm DM, Wax LM, Stoller EW (2000) Weed suppression for weed management in corn (Zea mays) and soybean (Glycine max) production systems. Weed Technol 14:713–717. https://doi.org/10.1614/0890-037X(2000)014[0713:WSFWMI]2.0.CO;2

Aria M, Cuccurullo C (2017) *bibliometrix*: An R-tool for comprehensive science mapping analysis. J Informet 11:959–975. https://doi.org/10.1016/j.joi.2017.08.007

Belfry K, Sikkema PH (2018) Weed management in adzuki bean: A review. Can J Plant Sci 98:1221–1233. https://doi.org/10.1139/cjps-2018-0026

Bond W, Grundy AC (2001) Non-chemical weed management in organic farming systems. Weed Res 41:383–405. https://doi.org/10.1046/j.1365-3180.2001.00246.x

Carrubba A, Calabrese I, Ascolillo V (2009) Non-chemical weeds management in two mediterranean culinary herbs. In: Acta Hortic. International Society for Horticultural Science, pp 51–58

Debaeke P, Perronne R, Colbach N et al (2024) Non-chemical weed management: Which crop functions and traits to improve through breeding? Crop Prot 179. https://doi.org/10.1016/j.cropro.2024.106631

Gazoulis I, Kanatas P, Papastylianou P et al (2021) Weed management practices to improve establishment of selected lignocellulosic crops. Energies 14. https://doi.org/10.3390/en14092478

Hatcher PE, Melander B (2003) Combining physical, cultural and biological methods: Prospects for integrated non-chemical weed management strategies. Weed Res 43:303–322. https://doi.org/10.1046/j.1365-3180.2003.00352.x

Kumar S, Kumari S, Rana SS et al (2024) Weed management challenges in modern agriculture: The role of environmental factors and fertilization strategies. Crop Prot 185. https://doi.org/10.1016/j.cropro.2024.106903

Li S, Feng Q, Yu S et al (2026) The application and challenges of physical technology in modern agricultural plant protection. Eur J Agron 174. https://doi.org/10.1016/j.eja.2025.127944

Mennan H, Jabran K, Zandstra BH, Pala F (2020) Non-chemical weed management in vegetables by using cover crops: A review. Agronomy 10. https://doi.org/10.3390/agronomy10020257

Migliorini P, Wezel A (2017) Converging and diverging principles and practices of organic agriculture regulations and agroecology. A Rev Agron Sustain Dev 37. https://doi.org/10.1007/s13593-017-0472-4
Monteiro A, Santos S (2022) Sustainable Approach to Weed Management: The Role of Precision Weed Management. Agronomy 12. https://doi.org/10.3390/agronomy12010118
Nawaz A, Farooq M, Nadeem F et al (2019) Rice-wheat cropping systems in South Asia: issues, options and opportunities. Crop Pasture Sci 70:395–427. https://doi.org/10.1071/CP18383
Ofosu R, Agyemang ED, Márton A et al (2023) Herbicide Resistance: Managing Weeds in a Changing World. Agronomy 13. https://doi.org/10.3390/agronomy13061595
Pannacci E, Lattanzi B, Tei F (2017) Non-chemical weed management strategies in minor crops: A review. Crop Prot 96:44–58. https://doi.org/10.1016/j.cropro.2017.01.012
Roberts J, Peerzada AM, Bajwa AA (2024) Biology, Ecology, Impacts and Management of the Invasive Weed, Blue Heliotrope (Heliotropium amplexicaule Vahl)—A Review. Sustainability 16. https://doi.org/10.3390/su16145923
Rodenburg J, Zwart SJ, Kiepe P et al (2014) Sustainable rice production in African inland valleys: Seizing regional potentials through local approaches. Agric Syst 123:1–11. https://doi.org/10.1016/j.agsy.2013.09.004
Scopus (2026) Scopus–Homepage.. Accessed March 28 2026. https://www.scopus.com/pages/home#basic
Shekhawat K, Rathore SS, Dass A et al (2017) Weed menace and management strategies for enhancing oilseed brassicas production in the Indian sub-continent: A review. Crop Prot 96:245–257. https://doi.org/10.1016/j.cropro.2017.02.017
Wei D, Liping C, Zhijun M et al (2010) Review of non-chemical weed management for green agriculture. Int J Agric Biol Eng 3:52–60. https://doi.org/10.3965/j.issn.1934-6344.2010.04.052-060

Chapter 3
Stakeholder's Engagement in Medicinal and Aromatic Plants Sector

Katarzyna Bączek, Zenon Węglarz, Ivanka Semerdjieva, Saadet Tuğrul Ay, Tadeusz Pęczek, Agnieszka Bądel, Paweł Sobczyk, and Artur Sobolewski

3.1 Introduction

The medicinal and aromatic plants (MAPs) sector constitutes a multifaceted and increasingly significant field situated at the intersection of biodiversity conservation, traditional knowledge systems, and emerging bio-economy dynamics.

This chapter examines the subject of stakeholders engagement in the sector of medicinal and aromatic plants. The analysis starts with the characteristics of value chains for the sector of MAPs. All the important value chain components will be explored, as well as actions directly supporting value chains. The chapter will then

K. Bączek · Z. Węglarz
Professor, Department of Vegetable and Medicinal Plants, Institute of Horticultural Sciences, Warsaw University of Life Sciences–SGGW, Warszawa, Poland
e-mail: katarzyna_baczek@sggw.edu.pl

Z. Węglarz
e-mail: zenon_weglarz@sggw.edu.pl

I. Semerdjieva
Associate Professor, Department of Botany and Agrometeorology, Agricultural University Plovdiv, Plovdiv, Bulgaria

Associate Professor, Department of Fungal Diversity and Resources, Institute of Biodiversity and Ecosystem Research, Bulgarian Academy of Sciences, Sofia, Bulgaria

I. Semerdjieva
e-mail: v_semerdjieva@abv.bg

S. T. Ay
Antalya Metropolitan Municipality, Climate Change Department, Clean Energy, Antalya, Türkiye
e-mail: tugrulay@hotmail.com

T. Pęczek · A. Bądel · P. Sobczyk · A. Sobolewski (✉)
Foundation for Education and Social Dialogue PRO CIVIS, Kielce, Poland
e-mail: a.sobolewski@procivis.org.pl

V. J. P. D. Martinho et al. (eds.), *Chemical-free Weed Control*,
SpringerBriefs in Agriculture, https://doi.org/10.1007/978-3-032-34699-5_3

move on to address networks and initiatives focused on non—chemical weed management in medicinal and aromatic plants in the EU and in the EU associated countries, outlining also the subject of policy advocacy of the networks. As the stakeholders in the MAPs sector are diverse, including farmers, collectors, processors, traders, retailers, policy makers, financial institutions, researchers, and consumers, there is a need to analyse the intersections and stakeholders' expectations along the value chains. The next subject of importance will be tools and methods with the potential to improve partners engagement in networks and partnerships. The general observations conclude with the results of the mapping of selected existing medicinal and aromatic plants' networks and their characteristics in terms of such features as co-operation, needs, limitations and expectations. Furthermore the medicinal and aromatic plants systems, networks and value chains in Poland, Türkyie and Bulgaria will be presented.

Structured stakeholder engagement emerges as a critical precondition for strengthening medicinal and aromatic plants value chains. The stakeholder networks offer an essential platform for aligning sectoral practices with broader socio-environmental objectives. This chapter therefore examines the architecture, functions, and cooperative modalities of medicinal and aromatic plants stakeholders and networks, providing analytical insights into how coordinated engagement can advance both the sustainability and competitiveness of the sector.

The following methodology has been employed for this chapter.

(1) Own research into:

- the value chain components of MAPs,
- existing MAPs international, national and regional networks,
- tools and methods improving partners engagement in networks and partnerships,
- the scope and principles of cooperation between partners in the MAPs networks.

The research was done based on available external sources, own observations, analysis of core documents, and was further backed by literature review.

(2) Questionaries/interview scenarios addressed to 18 medicinal and aromatic plants networks from 7 countries, i.e. Bosnia & Herzegovina, Bulgaria, Italy, Kosovo, Poland, Serbia and Türkiye. The scenarios inquiring in particular about the cooperation (with other networks, business partners, universities, public authorities, research units), needs and limitations of the network, communication and monitoring, expectations and recommendations.

T. Pęczek
e-mail: t.peczek@procivis.org.pl

A. Bądel
e-mail: a.badel@procivis.org.pl

P. Sobczyk
e-mail: p.sobczyk@procivis.org.pl

3.2 The Characteristics of Value Chains for the Sector of Medicinal and Aromatic Plants

The value chain for medicinal and aromatic plants encompasses all the steps from the cultivation and wild collection of plant materials to the processing, manufacturing, and marketing of finished products used in medicinal, cosmetic, food, and perfumery industries. The MAPs value chain is crucial for sustainable development in rural areas and requires strong collaboration across stakeholders to ensure product quality, fair income distribution, and market access. Well organized MAPs value chains have the following important characteristics:

- "diversity", resulting from a wide variety of plants, processing methods, and end-products; as "interconnectedness", due to the fact that MAPs serve as raw or intermediate inputs for several high-value sectors, including cosmetics, pharmaceuticals, and food processing;
- "variability", resulting from the dependence on proper harvesting, drying, processing, extraction, and storage to maintain quality and marketability;
- "fragmentation", due to many small processors and inconsistent long-term trading relationships.

In terms of opportunities MAPs offer potential for rural employment, export income, and women's empowerment. At the same time MAPs value chains face constraints such as quality assurance, asymmetric market information, and sustainability issues.

In the area of MAPs production—value chain stages depend on four types of primary raw materials, i.e. fresh plants material, dried plant material, essential oils and extracts.

To properly identify the stakeholders active in the MAPs sector there is a need to properly assess value chain components, as presented in the Table 3.1.

Beside the basic stages of MAPs value chain and related key activities as presented in the Table 3.1, there is also the need for well-developed actions directly supporting value chains like logistics, transport, brokerage, physicochemical analyses, and regulatory control. Furtheromore MAPs value chains have to be assisted by:

- public sector entities representing the EU level institutions, national governments, state agencies, control and standardization units, regional and local authorities;
- entities supporting sector development, in particular education, academia, research, financial sector (EU and national public funds, banks, venture capital firms, business angels, insurance companies), technology parks, entrepreneurship incubators, clusters, technology HUBs, chambers of commerce, branch associations, producers' cooperatives, public agriculture advisory services and private consulting (Chandra et al. 2021).

Strengthening the MAPs value chain requires addressing quality standards, building capacity among producers, improving market and price information systems, promoting sustainable harvesting/cultivation, and supporting gender and social equity throughout the chain.

Table 3.1 MAPs value chain components (EU CAP 2025)

	Value chain stage	Main actors	Key activities
1	Input supply	Propagators, input suppliers	Production and provision of seeds, seedlings, and other propagation materials, as well as farming inputs like fertilizers, agrochemicals, and machinery
2	Cultivation/ Collection	Farmers, collectors	Medicinal and aromatic plants are either cultivated on farms or collected from the wild by collectors and farmers
3	Postharvest handling	Postharvest services suppliers, storage	Includes primary processing such as cutting, cleaning, drying, and storage of raw plant materials to preserve quality
4	Primary processing	Processors	Further steps like sorting, grinding, and packaging of dried plant materials for bulk sale or preparation for extraction
5	Secondary processing	Extractors, biorefineries, labs	Extraction and purification processes to obtain essential oils, oleoresins, extracts, or other value-added plant products; these often require specialized technical skills and facilities
6	Manufacturing and formulation	Manufacturers	Development and formulation of final products (such as herbal medicines, dietary supplements, cosmetics, and fragrances) from MAPs-based extracts and raw materials
7	Certification/ Quality control	Labs, certifiers	Analytical laboratories perform quality, safety, and regulatory checks. Certification for organic, sustainable, and other standards is often provided at this stage
8	Distribution/Trade	Distributors, wholesalers	Bulk and packaged products move through various marketing channels to wholesalers, retailers, and exporters before reaching end-users
9	End-use	Retailers, gastronomy, HoReCa	Products are finally sold to consumers through pharmacies, herbalists, organic stores, and supermarkets, or further processed in other industries (pharma, perfumery, food)

3.3 MAPs' Networks, Their Members and Practises of Cooperation

Main approach of this section is to address networks and initiatives focused on non—chemical weed management (NCWM) in medicinal and aromatic plants in the EU and in the EU associated countries. Most of the networks and initiatives have been established because of EU financed projects and are aiming to promote MAPs cultivation, processing, and trade. These networks often connect researchers, farmers, and businesses to share knowledge and best practices, fostering sustainable development and innovation within the MAPs sector.

MAPs networks represent a robust alliance of science, policy, and enterprise—universities and research centers, government agencies, grower and industry associations, trading and processing businesses—each providing critical pools of specialized knowledge and practical expertise. Their close collaboration ensures continual flow of best practices, innovations, market intelligence, and strategic guidance across the national and international MAPs value chain.

3.3.1 Range of Partners

1. Scientific Partners

   - universities & research institutes—leading universities (e.g. University of Pisa UNIPI, Italy, Peruvian and Brazilian research labs) and national institutes (such as the Institute of Natural Fibres and Medicinal Plants (IWNiRZ) in Poland, and the National Institute of Medicinal and Aromatic Plants, Morocco) are core to research, genetic resources conservation, bioactivity studies, and innovation;
   - international scientific networks & projects—which assure multinational collaborations like the EMAP Project (with 14 partner institutions from the EU and other countries), ECPGR Medicinal and Aromatic Plants Working Group (European Cooperative 2025) collaborative platforms such as Nord-BaltMAP (for seed production knowledge exchange), and research clusters in the Amazon and Europe which foster transnational sharing of best practices, innovations, and data.

2. Public Partners

   - government agencies & regulatory bodies—ministries of agriculture, health, environment, and dedicated public agencies (e.g. Morocco's National Agency for Medicinal and Aromatic Plants, ANPMA) which support sector policy, certification, public investment, sustainability, and regulatory harmonization across nations;
   - international organizations—groups such as the IUCN SSC Medicinal Plant Specialist Group (MPSG) acting globally for conservation, sustainable use, and policy development, often partnering with national governments and academic centers (IUCN SSC 2025);
   - funding bodies & rural development initiatives—European Agricultural Fund for Rural Development and national rural innovation programs (like Portugal's MAP initiative) underpin national and cross-border MAP networks focused on research, training, and sectoral development.

3. Business Partners

- producers & farmers—national and regional MAPs grower associations, cooperatives, and networks of wild plant collectors, which serve as the foundation of sector knowledge, linking practical cultivation know-how with new technical advances;
- processing & trading companies—entities such as Herbapol in Poland, and national consortia of processors and exporters in other MAPs-producing countries, which bridge the gap between growers and global buyers, and facilitate application of standards and market-oriented innovations;
- sector organizations & clusters—platforms including EUROPAM (European Federation of Essential Oils and MAPs), national business consortia, and market-driven public–private partnerships (like MAP 2ERA for Morocco–EU collaboration), which drive innovation, certification, best practice adoption, and value chain upgrading;
- knowledge-sharing & stakeholder platforms—recent initiatives (such as India's proposed National Stakeholder Platform for Sustainable Medicinal and Aromatic Plants and various digital traceability and sustainability tool pilots), which provide direct engagement channels for business operators and facilitate exchange with both scientific and public partners.

Through these intertwined networks, direct pools of applied and theoretical knowledge are mobilized. Notable mechanisms include:

- researcher and staff exchanges between scientific institutions and industry;
- joint research programs and collaborative field projects involving universities, government agencies, and private sector actors;
- national and international training events, knowledge hubs, and expert-driven technical publications accessible to practitioners and businesses;
- digital tools and traceability platforms that connect growers, processors, traders, and policymakers for real-time information flow and joint problem solving;
- stakeholder roundtables, webinars, and policy development workshops open to all sector representatives.

3.3.2 *Existing Medicinal and Aromatic Plants International and National Networks*

Several established international networks and associations focus on MAPs, working to advance research, conservation, sustainable use, and trade. Key organizations and collaborative structures include:

- International Council for Medicinal and Aromatic Plants (ICMAP) is a major global body that promotes international collaboration, scientific exchange, and organizes world congresses on MAPs. ICMAP brings together scientific, governmental, and industry stakeholders to address MAPs issues worldwide (International Council 2025).

- Society for Medicinal Plant and Natural Product Research (GA), which is primarily research-focused. GA connects professionals globally and supports research dissemination, especially through its publications and events (Society 2025)
- ECPGR Medicinal and Aromatic Plants Working Group is part of the European Cooperative Programme for Plant Genetic Resources (ECPGR) coordinating conservation, research, and uses of MAPs in Europe, and focusing on gene banks, biodiversity, and sustainable sourcing.
- EUROPAM European Herbs Growers Association represents European national associations and provides expertise on legislation, best practices (including good agriculture and collection practices), and fosters collaboration among producers and stakeholders on the continent (Europam 2025).
- Association for Medicinal and Aromatic Plants of Southeast European Countries (AMAPSEEC) is the regional body enhancing cooperation among Southeastern European nations, sharing information, organizing events, and advancing sustainable cultivation and trade (Amapseec 2025).
- Association of Medicinal and Aromatic Plants of the Mediterranean (AMAPMED) supports collaboration among Mediterranean organizations, often working with AMAPSEEC and the Global Federation of Medicinal and Aromatic Plants.
- Conferences and Symposia–recurrent global summits (e.g. the Global Summit on Medicinal and Aromatic Plants) and international symposia offer essential networking, partnerships, and knowledge-exchange opportunities among researchers, practitioners, NGOs, policymakers, and industry.

These networks facilitate the sharing of research, best practices, and confer collective influence on international policy, trade, and conservation priorities for MAPs. They also provide platforms for ongoing coordination and innovation across regions and sectors, helping to ensure both the sustainability of wild resources and the economic viability of MAP value chains.

Apart from the international networks there are obviously also the national MAPs networks or associations. These may include plant growers' associations, herb producer cooperatives, botanical research networks, or traditional medicine organizations. Examples include:

- Bulgaria–active national associations and a large sector for MAP cultivation and processing.
- France–"Syndicat des Simples" and several regional groups linked to medicinal and aromatic plants.
- Poland–national research centres and producers' associations coordinated efforts for export and quality standards.
- Germany, Hungary, Italy, Romania, and Spain—each has a dedicated sector organization or national network focusing on research, cultivation, processing, and export of MAPs, often tied to national ministries or agricultural chambers.

For up-to-date contacts and a detailed list of national representatives, the ECPGR MAP Working Group website maintains the current network of MAPs experts, organizations, and delegate contacts for each participating country. Sector ministries (Agriculture/Health), universities' agricultural faculties, and national plant genetic resource programs are also strong entry points to find relevant national MAPs networks and associations.

3.3.3 Policy Advocacy of the Networks

Policy advocacy is a core function of MAPs stakeholder groups and networks-enabling collective action to shape supportive legislation, funding, standards, and sectoral priorities at both national and international levels. The policy advocacy is carried out by utilizing following mechanisms:

1. Sector representation and unified voice

Networks such as the Polish Herbal Committee (PKZ), Institute of Natural Fibres and Medicinal Plants (IWNiRZ), EUROPAM, and ECPGR act as the unified voice for MAPs stakeholders (farmers, processors, researchers, and traders) in policy consultations with government ministries, the EU, and international agencies. These bodies aggregate the needs and perspectives of diverse actors, ensuring that rural producers, processors, and others are not marginalized in public policy.

2. Advocacy objectives and activities

Groups lobby for favourable laws concerning sustainable harvesting, seed certification, good agricultural and collection practices (GACP), organic standards, and fair trade that protect both industry and biodiversity. Stakeholder networks advocate also for financial incentives, research grants, and rural development programs targeting MAPs. Furthermore working with trade authorities—advocacy efforts focus on reducing export barriers, developing traceability systems, and negotiating for inclusion in free trade agreements.

3. Mechanisms and means of advocacy

Networks develop detailed policy recommendations and white papers submitted to lawmakers and ministries, reflecting consensus positions from the widest array of stakeholders. They organize roundtables, working groups, and dialogues where stakeholders and policymakers interact directly, fostering co-creation of sectoral policy. Moreover the networks function as representation in official policy processes (e.g. European Parliament consultations, national policy review boards) ensuring sector involvement in drafting legal standards or subsidy schemes. They engage also in awareness-raising among politicians, the media, and the public about the social, economic, and environmental benefits of MAPs, building broader support for their policy goals.

4. Advocacy outcomes and achievements

Many stakeholder-driven campaigns have led to adoption of good collection and manufacturing practices, support for MAPs gene banks, and recognition of MAPs as a strategic sector in rural policy. Advocacy has resulted also in more favourable export conditions, funding for innovation, and national policies promoting MAPs in organic agriculture. Efforts have led to formal recognition of the roles of smallholders, indigenous collectors, and women in the MAPs value chain, with policies supporting equity and fair benefit sharing.

As presented—stakeholder networks in the MAPs sector play a decisive role in collective advocacy by representing the value chain in policymaking, shaping legislative and regulatory agendas, securing support for sustainability and growth, and promoting an enabling environment for all actors. This advocacy is built on inclusive participation, evidence-based recommendations, and a commitment to the long-term viability of MAPs resources and enterprise.

3.3.4 The Scope of Cooperation Between Partners in the MAPs Networks

The scope of cooperation between partners in medicinal and aromatic plants networks is broad and structured to maximize knowledge exchange, innovation, sustainability, and economic success across scientific, public, and business spheres. This cooperation embodies the following activities:

1. Research and innovation, where partners (universities, research institutes, businesses) jointly conduct R&D projects targeting new MAPs varieties, sustainable cultivation, quality assurance, phytochemistry, genetics, and product innovation. Activities may include collaborative grant applications, shared laboratory/field trials, and exchange of technical expertise.
2. Genetic resource conservation and biodiversity, with cooperation on conserving genetic diversity through national seed banks, coordinated *ex situ* and *in situ* conservation programs, and joint exploration/collection missions. Networks like ECPGR facilitate standardized protocols and access to plant material for all member countries.
3. Policy development & standardization, where partners collaborate with government agencies and regulatory bodies to shape sector policy, set quality and safety standards (e.g. good agricultural and collection practices), and harmonize regulations.
4. Capacity building & training, with the organization of joint training sessions, knowledge hubs, workshops, webinars, and field days for farmers, processors, and researchers. Exchanges of expertise and best practices may also happen via study visits, internships, and secondments.
5. Marketing, processing & trade, where coordinated efforts to enhance market access, fair trade, traceability, and promotion of MAPs products takes place.

Business clusters, cooperatives, and processors shall align on shared branding and value-adding strategies.
6. Sustainable cultivation & conservation, with the promotion of sustainable wild collection practices and transition to cultivation to reduce overharvesting. Partners cooperate on certification, monitoring, and compliance with international sustainability standards.
7. Information & digital platform sharing, with collaborative development and use of digital traceability tools, data-sharing platforms, and sectoral dashboards to facilitate real-time knowledge transfer and transparency.

3.4 Stakeholders' Needs and Expectations in MAPs Sector

As it was already pointed out the stakeholders in the medicinal and aromatic plants sector are diverse-farmers, collectors, processors, traders, retailers, policy makers, financial institutions, researchers, and consumers—each with distinct needs and expectations along the value chain. Understanding these needs and expectations is crucial for effective sector development, sustainability, and value delivery (see Table 3.2).

There are also cross—cutting features like collaboration and coordination, sustainability, capacity building and training, digitalization and traceability.

3.5 Tools and Methods to Improve Partners Engagement in Networks and Partnerships

To improve partner engagement in networks and partnerships, a combination of technology tools and strategic methods is essential. Modern approaches focus on streamlining communication, fostering collaboration, aligning goals, and providing real value to partners. It must be however noticed that these tools are of general nature and are not confined only to the MAPs sector.

The following essential tools must be named as leading to improving partners engagement in networks and partnerships:

- Partner Relationship Management Platforms (PRM): these Platforms centralize partner onboarding, resource sharing, deal registration, rewards, and performance tracking. Leading platforms (such as Kiflo, Introw, Impartner, and Zift Solutions) offer real-time analytics, automated communications, and easy integration with partners' existing workflows.
- Collaboration Tools: Platforms like Slack, Teams, and shared document environments facilitate seamless interaction, updates, and file sharing–meeting partners where they already work and cutting friction.

Table 3.2 Needs and expectations of the MAPs stakeholders

Type of stakeholder	Needs	Expectations
Farmers, collectors and producers	(1) Access to high-quality seeds/planting materials, fertilizers, and organic inputs (2) technical support, training, and capacity building (especially for sustainable cultivation, wild collection, and value addition) (3) Fair and transparent market access, stable and remunerative prices, and support with contracts or buy-back arrangements (4) Access to infrastructure for drying, sorting, storage, and transportation	(1) Policy and subsidy support for both traditional and modern MAP production (2) Reduced market risk, assured buyers, and minimized exploitation by middlemen (3) Inclusion in decision-making networks, cooperatives, and sector platforms
Processors and Industry	(1) Consistent supply of raw material with standardized quality and traceability (2) Access to modern processing technology and infrastructure (drying, extraction, distillation) (3) Skilled labour for processing and value-added product development (4) Clarity and support on quality, safety, and certification standards	(1) Collaboration with growers for contract farming or direct sourcing (2) Supportive policy and regulatory environment (3) Access to export markets and assistance in meeting international certification and traceability
Traders, wholesalers and retailers	(1) Reliable supply chains and predictable delivery schedules (2) Information about market trends, prices, and consumer preferences (3) Assurance of legal compliance (CITES, phytosanitary, food safety)	(1) Cooperative marketing approaches and branding (2) Efficient infrastructure for storage, handling, and logistics (3) Stronger sectoral representation with policymakers and regulatory bodies

(continued)

Table 3.2 (continued)

Type of stakeholder	Needs	Expectations
Consumers	(1) Safe, high-quality, and traceable medicinal and aromatic plant products (2) Accurate product information (origin, certification, health benefits) (3) Increasing demand for sustainably sourced, organic, and ethically produced MAP goods	(1) Access to a variety of traditional and new products (2) Reasonable pricing, clear labelling, and consumer protection
Policy makers and regulatory bodies	(1) Sector data for informed policy and regulation (2) Input from all stakeholders to shape inclusive, practical policies (3) Tools to monitor sustainable use and prevent overharvesting or illegal trade	(1) Enhanced stakeholder coordination for effective implementation (2) Adoption of standards for Good Agricultural & Collection Practices, traceability (3) Sector contributions to rural development, health, and export growth
Researchers, Academic and Extension Services	(1) Resources for scientific innovation, breeding, and value chain research (2) Platforms for knowledge sharing, stakeholder engagement, and technology transfer	(1) Sector support for R&D priorities (e.g., conservation, phytochemistry, climate resilience) (2) Uptake of research results by industry and policymakers

- Feedback & Survey Portals: Regular partner satisfaction surveys, suggestion boxes, and feedback tools empower partners to voice concerns and drive improvements.
- Performance Dashboards: Real-time dashboards help tracking progress, engagement, and KPIs for transparent, data-driven decision-making.

There is also a catalogue of already proven methods & best practices improving partners engagement in networks and partnerships. These include in particular:

- personalized engagement, which segments partners and deliver communication, training, and resources tailored to their specific needs and goals;
- regular, transparent communication with scheduled consistent check-ins, updates, newsletters, and feedback loops, all of which shall empower partners to stay informed and involved;
- joint business planning collaborating on goal setting, strategy, and execution to ensure shared objectives and mutual benefit, fostering a true sense of partnership;

- interactive training & enablement, which offer live webinars, on-demand onboarding, and ongoing education to help partners grow skills and confidence;
- systems of recognition & incentives, which celebrate partner achievements with public recognition, awards, gamification, or financial rewards to reinforce engagement and loyalty;
- easy access to resources providing intuitive portals or distributed marketing platforms with up-to-date content, co-brandable assets, and campaign support;
- open, collaborative culture, which encourages knowledge sharing, networking (through partner advisory boards or social groups), and innovation challenges to build community and trust.

Furthermore in order to track the approach and maximize impact—measurements and optimization methods such as track tool usage, engagement, partner-driven revenue, and retention rates could be employed—to continually refine the approach and maximize impact.

Out of the presented catalogue of essential tools feedback portals need special attention. They play a crucial role in strengthening long-term partner commitment and loyalty by creating a culture of open communication, continuous improvement, and responsiveness. When partners can share feedback and see their suggestions result in real changes, they feel heard and valued. This sense of inclusion builds deeper trust and emotional investment in the partnership, reinforcing commitment over time. Feedback portals enable companies to quickly identify and remedy issues that could jeopardize partner satisfaction. Proactively addressing concerns signals genuine care, reducing frustration and risk of defection, and ensuring partners remain loyal. Well-managed portals allow for tailored support, resources, and communications based on partner feedback. This creates more relevant, engaging experiences that strengthen satisfaction and reinforce the desire to continue the relationship. Analytics from feedback portals help organizations understand partner needs and preferences, enabling ongoing enhancements to programs and communication channels. This results in a partnership that evolves and adapts, further securing loyalty. By sharing how feedback will be used and closing the loop with updates on resulting actions—organizations build transparency. Portals with built-in reward, recognition, or gamification elements linked to feedback can further motivate partners, increasing participation and emotional attachment to the partnership.

Effective feedback portals transform transactional partnerships into dynamic, trust-based collaborations. By listening to partner voices, acting on their input, and continually optimizing the partner experience—organizations not only increase satisfaction and engagement, but foster also the strong commitment and loyalty necessary for enduring, mutually rewarding relationships.

A significant and positive impact on partner motivation levels in networks and partnerships is being played by collaboration and communication tools. They exercise in particular the following features:

- foster a sense of belonging and community—real-time chats, shared project spaces, and access to centralized resources help partners feel connected and part

of a team, even if they are geographically dispersed; this sense of belonging boosts motivation by reinforcing that each partner's contributions matter;
- improve transparency and trust—these tools facilitate open, clear communication and resource sharing and are building trust among partners; being able to access up-to-date information, track progress, and voice feedback in real time ensures partners feel informed and empowered, which is strongly linked to higher engagement and commitment;
- enable recognition and inclusion—collaboration platforms allow for public acknowledgments, shoutouts, and shared successes; when achievements are visible across the network, partners feel valued and respected, increasing their desire to go the extra mile;
- reduce friction and barriers—easy-to-use communication and collaboration tools streamline managing joint tasks, scheduling, and information exchange; this reduces delays, misunderstandings, and administrative burdens–allowing partners to focus on higher-value activities, which is inherently motivating;
- strengthen relationships and culture—frequent, structured, and informal exchanges (like quick check-ins or group brainstorming) nurture stronger interpersonal relationships, making teams more cohesive and motivated toward shared goals; stronger relationships often correlate with higher levels of intrinsic motivation;
- acelerate learning and innovation—by making it simple to share knowledge, provide feedback, and collaborate on new ideas, these platforms enhance opportunities for personal and professional growth, fuelling directly the motivation;
- provide actionable insights—todays advanced platforms offer analytics on activity, participation, and engagement; these insights help organizations identify the most engaged partners and design targeted motivation or support strategies.

When properly implemented, collaboration and communication tools not only improve operational efficiency but are also a powerful lever for partner motivation, driving engagement, loyalty, and shared success in partnership ecosystems. A robust communication plan for the continual identification and engagement of stakeholder groups and networks in MAPs sector should be structured. In order to obtain such a plan for the MAPs the following substantive and operational areas must be observed: (1) conducting regular stakeholder analysis, (2) maintaining a living stakeholder database, (3) using of multi-channel communication, (4) monitoring the network, (5) leveraging partnerships, (6) utilising continuous engagement cycles, (7) conducting monitoring, evaluation and adaptation activities.

3.6 The Networks and Its Stakeholders—Practical Functional Approach

Within the scope of the COST Action "Non-chemical weed management in medicinal and aromatic plants (MAPs)" (weedingMAPs) the project team conducted a systematic mapping of selected existing MAPs networks using a standard questionnaire template. In total, information on 18 networks across seven countries was collected, providing valuable insight into their structure, cooperation, and needs. The responses enabled the analysis summarised below, highlighting common challenges, opportunities, and recommendations for strengthening the MAP sector.

The analysed networks represent a diverse landscape of entities involved in the MAPs sector. Most of them are formally registered organisations, including associations, cooperatives, and limited liability companies, while a smaller group consists of informal or academic initiatives. The networks typically bring together producers, processors, researchers, and other stakeholders along the MAPs value chain. Their organisational maturity varies from long-established national associations to newly formed business clusters or company-based partnerships operating under corporate structures.

In terms of objectives, most networks aim to support sustainable MAPs production, improve cooperation between science and business, ensure product quality and certification, and strengthen the competitiveness of the sector. Several networks also engage in policy dialogue and advocacy, contributing to the development of a favourable legal and institutional environment for MAP producers and processors.

Only a limited number of networks explicitly address NCWM. These typically integrate mechanical or manual weed control methods as part of their organic or environmentally friendly cultivation practices. The majority do not mention NCWM directly but demonstrate growing awareness and openness towards sustainable and low-input production methods, which could provide a foundation for wider NCWM adoption in the future.

Co-operation

Most networks maintain active cooperation with a variety of external partners, including research institutions, universities, public authorities, donor organisations, and private companies operating along the MAPs value chain. This cooperation primarily aims to enhance knowledge transfer, innovation, and market access, as well as the participation in joint projects supporting organic certification, sustainable harvesting, and product promotion. Some networks, particularly business-led ones, maintain strong ties with foreign buyers and certification bodies, while academic and association-based networks focus more on capacity building and sectoral policy development.

The scope and intensity of cooperation depend on organisational capacity and resources. Larger entities and those integrated with private companies often sustain international partnerships and project-based collaboration, while smaller or informal networks cooperate mainly at the local or national level. A number of networks plan to expand their cooperation towards regional initiatives or EU-funded programmes, recognising the added value of transnational networking and experience exchange. Overall, collaboration is seen as a crucial factor in strengthening competitiveness and aligning practices with EU standards in the MAPs sector.

Needs and limitations

As for the needs and limitations across all networks–the most frequently reported needs concern human resources, education, access to finance, and legal clarity. Many networks struggle with limited access to trained staff, lack of specialised expertise in organic production or in certification, and insufficient funding to expand activities. Education and training remain a key demand, particularly for younger producers and new entrants to the MAPs value chain. Access to sectoral data and research results is also limited, hindering evidence-based decision-making and innovation.

Several networks point to complex administrative procedures, weak institutional support, and absence of dedicated national policies for the MAPs sector. Financial constraints are a recurring challenge, especially for small producers and associations that rely on membership fees or donor support. Despite these barriers, many networks express a strong willingness to improve knowledge transfer, mobility, and cross-sectoral cooperation. The overall picture indicates a sector with high potential but constrained by limited resources and regulatory obstacles, requiring targeted public and institutional support.

Communication and monitoring

Communication within the networks is carried out mainly through email, online platforms, and regular meetings or workshops, ensuring information flow among members. Larger or business-led networks use structured tools such as digital databases, social media, or reporting systems, while smaller associations rely more on direct or informal exchanges. In most cases, information shared covers policy updates, training opportunities, and market developments, with varying levels of transparency and frequency depending on organisational capacity.

Monitoring of members activities remains uneven across networks. Some entities, especially companies and well-established associations, track participation, compliance with standards, and contributions to joint initiatives, while others lack formal monitoring systems. New development opportunities are often shared through workshops or partner consultations. Overall, most networks maintain an open and transparent communication culture, although few have institutionalised mechanisms for systematic performance monitoring or impact evaluation.

Expectation and recommendations

The majority of networks express expectations for stronger institutional and financial support for the MAPs sector at both national and EU levels. Key recommendations include simplifying administrative procedures, harmonising legislation with EU standards, and establishing dedicated subsidy schemes for MAPs producers. Networks emphasise the need for improved access to funding, investment in modern equipment and technologies, and enhanced cooperation between science, business, and public authorities.

Several participants also recommend promoting research, innovation, and training in sustainable and organic production, including NCWM. Developing consistent quality standards and certification systems is seen as essential to increase the competitiveness of MAPs products on international markets. Networks call for more visibility of the sector, stronger public-private partnerships, and the replication of successful good practices and case studies across countries. Overall, the sector seeks a coordinated policy approach that recognises its strategic role in rural development, biodiversity conservation, and the green transition.

3.7 Countries Case Studies

3.7.1 Poland

3.7.1.1 Polish MAPs System

Poland is one of the largest producers of herbal raw materials in Europe. These products are collected from both wild-growing and cultivated plants. It is estimated that about 200 wild-growing plant species are commercially harvested. The largest quantities include horse chestnut seeds, elderberry fruits and flowers, nettle leaves, and birch leaves. These raw materials are obtained mainly from the eastern and southern parts of Poland. Despite its strong export position, the Polish wild collection system faces challenges, including sparse and inconsistent legislation and limited monitoring of harvesting practices. The supply chain also shows a high degree of informality and lacks transparency, raising concerns about long-term sustainability and the protection of vulnerable plant species. Rural collectors continue to rely on MAP harvesting as a supplementary source of income, although profitability has declined in recent years due to market changes. Such activity is primarily undertaken by older people living in rural areas and nowadays is considered less attractive to the younger generation.

At the same time, global demand for natural products, herbal medicines, and essential oils continues to rise, supporting ongoing commercial interest in Polish MAPs resources. Sustainable cultivation and improved processing practices are increasingly emphasized as pathways to strengthen Poland's position in the growing international market for MAPs raw materials. The most cultivated species in Poland are thyme, peppermint, valerian, purple coneflower, milk thistle, savory, lovage, marshmallow, and common chamomile. The currently observed trend also involves the introduction of wild-growing species into cultivation. Herb plantation areas vary greatly in the country, ranging from several hundred square meters to several hundred hectares. It is worth emphasizing that due to the need for crop rotation, herbal farms cultivate not only MAPs but also typical agricultural species. In Poland, there are clear regional differences in the area of herb cultivation. The biggest MAPs farms are situated in western and central Poland (Wielkopolska, Kujawy and Mazowsze), whereas the smallest–in the eastern part of the country (Lubelszczyzna). It is estimated that approximately 25,000 tons of dry herbal raw materials are obtained annually from cultivation, however, significant year-to-year differences are observed Sucholas et al. (2021).

3.7.1.2 Medicinal and Aromatic Plants Networks in Poland

Poland has a network for medicinal and aromatic plants production consisting of: (1) farmers and herb collectors, (2) companies processing MAPs, (3) agricultural advisory centers and state supervisory bodies, (4) research institutes and agricultural universities, (5) industry-oriented organizations.

1. Farmers and herb collectors

Most herbal raw materials in Poland originate from controlled cultivation (in accordance with the GACP requirements). Part of this production is carried out using organic farming methods. It is estimated (data from the Polish Agency of Restructuring and Modernisation of Agriculture, Polish Chief Inspectorate of Commercial Quality of Agricultural and Food Products, as well as own research data) that the annual area covered by herbal crops in Poland is 20–25 thousand hectares, including organic crops covering approximately 2–3 thousand hectares. The farmers cooperate with diverse institutions and companies. However, the main source of means of production and knowledge on methods of cultivation is passed on to farmers by the managers-instructors hired by the companies that buy and process MAPs' raw materials. Those people usually cooperate closely with farmers through the whole cultivation process, on the basis of contracts—often established on long-term cooperation and trust. MAPs production is subject to control by a number of state regulatory bodies, including those that provide subsidies to farmers.

The collection of wild-growing MAPs in Poland is carried out mainly by elderly people, who deliver the raw materials directly to the collection points, located in villages, from which they are transported to companies processing MAPs raw materials. Currently, herb collectors deliver herbal raw materials in fresh and dried form. Drying is also performed at the collection points or at the herbal company. The farmers produce also MAPs seeds and seedlings, both for their own needs and for sale.

2. Major Polish herbal companies

The main Polish herbal companies are:

- "Herbapol" companies located in selected Polish regions—Herbapol Lublin Sp. z o.o., Herbapol Poznań (PZZ Herbapol S.A.), Herbapol Wrocław (WZZ Herbapol S.A.), Herbapol Kraków (KZZ Herbapol S.A.)
- FZL Sp. z o.o.

- Runo Sp. z o.o.
- Dary Natury Sp. z o.o.
- KAWON–HURT Nowak Sp. j.
- Labofarm Sp. z o.o.
- Colfarm S.A.
- Martin Bauer Polska Sp. z o.o.

The companies work as medium—or large-scale MAPs producers, processors, marketers, importers, and exporters. They cooperate closely with MAPs collectors and farmers, building their own networks of herbal raw materials suppliers, usually based on long-term cooperation and trust. In this system, herb production is based on contracting. Herbal companies provide trainings and individual consultations for farmers on cultivation methods, including plant protection, harvesting, and post-harvest processing. In some cases, they supply farmers with means of production (seeds, seedlings, packaging) and monitor all stages of cultivation and analyse each MAPs batches for compliance with quality standards. This will be done with the use of services of specialized analytical laboratories or their own laboratories. The companies offer additional services to other producers, including advanced drying, multi-stage cleaning, and MAPs packaging, improving their quality. The herbal raw materials produced in Poland are mainly exported to Germany, France, Italy, and the USA. In Poland, they are mainly used to produce herbal teas, spices, and increasingly extracts, as well as traditional herbal medicines (mainly galenic preparations) and cosmetics. It is worth noting that in recent years, farmers increasingly process the raw materials they obtain into simple products (jams, juices, herbal teas, spices, and packaged herbs) and sell them on their own farms or at local markets as part of agricultural retail trade.

3. Agricultural advisory centres and sate supervisory bodies

MAPs producers, including farmers, wild-growing plants collectors, manufacturers, and traders are subject to an extensive quality control system. This applies especially to herbal raw materials intended for the production of herbal medicines and/or cosmetics. State regulatory bodies, including those that provide subsidies to farmers, are presented in Table 3.3—the Polish MAP system.

In Poland, there is also a support system for farmers organized within the framework of the Agricultural Advisory Centre (CDR)—the central state unit of the Polish Ministry of Agriculture and Rural Development. The main aim of the CDR's activities is to improve the knowledge and skills of agricultural advisers as well as to develop and standardize the services provided to Polish farmers. CDR holds a legal status of an autonomous public institution. CDR as an umbrella of the national body cooperates with 16 Regional Agricultural Advisory Centres covering all territory of Poland, employing around 3500 field advisors. CDR is the hub stimulating professional development of Polish agricultural and rural advisors, setting directions for development of modern agriculture, development of farm—based economy, as well as implementation of new socio-economic solutions for rural business (Agriculture,

Table 3.3 Polish MAPs system

Institution, organization, network	Type	Focus areas (from the point of view of MAPs producers)	Connectivity and status
The agency for restructuring and modernization of agriculture (ARMA)	Governmental–an accredited paying agency	To support agriculture and rural development in Poland, at the level of MAPs production, it provides subsidies for farmers, verifies compliance of crops and applied treatments with the plan included in the application for subsidies, and controls the quality of products on the level of farms	Co-operations with the Ministry of Agriculture and Rural Development; as a spender of public funds, ARMA falls under the supervision of the Ministry of Finance
Agricultural advisory centres (16 regional agricultural advisory centres)	Governmental–working within the framework of the polish Ministry of Agriculture and Rural Development	Knowledge transfer aimed at agriculture and rural development; organization of workshops and trainings for farmers concerning e.g., the newest regulations and trends in crop production; assistance for farmers in preparing and submitting applications for subsidies from ARMA	Cooperates with central government and local authorities; farmers and producers organizations, rural NGOs, local communities, institutes, and agricultural universities
State sanitary inspection	Governmental–working within the framework of the polish Ministry of Health	Controls the methods of drying MAPs intended for food; supervises the production, processing, and trade of MAPs within the agricultural retail trade; controls the level of pyrrolizidine and tropane alkaloids in herbal teas; supervises the herb collection points	Cooperates with MAPs producers—Farmers, herb collectors, collection points, and herbal companies
Certification bodies (CBs)	Private companies that carry out tasks delegated to them by the inspection of the commercial quality of agricultural and food products	Authorized to carry out inspection and certification in the organic farming system, directly on MAPs farms, during their processing and trade	Services for organic farmers, herb collectors, herb collection points, companies, and shops

(continued)

Table 3.3 (continued)

Institution, organization, network	Type	Focus areas (from the point of view of MAPs producers)	Connectivity and status
Inspection of the commercial quality of agricultural and food products	Governmental–working within the framework of the polish Ministry of Agriculture and Rural Development	Supervises the commercial quality of Agri-food products in production and trade; Authorized to carry out inspections of the certification bodies (also to inspect the quality of controls done by CBs on farms)	Cooperation with CBs, MAPs organic farmers, herb collectors, herb collection points, companies, and traders
State veterinary inspection	Governmental–working within the framework of the polish Ministry of Agriculture and Rural Development	Controls the quality of MAPs intended for feed production	Cooperates with MAPs producers—Farmers, and companies
State plant health and seed inspection service	Governmental–working within the framework of the polish Ministry of Agriculture and Rural Development	Increasing knowledge of plant protection and seed production among growers and the agriculture sector; at the level of MAPs farms—The service inspects, and approves the sprayers used on plantations	Cooperates with the Ministry of Agriculture and Rural Development, farmers, producers, organizations, institutes, and agricultural universities
Local government organizations	Governmental	Granting excise tax refunds for fuel used in agricultural production; granting permits for the construction/demolition of agricultural buildings; granting subsidies for agricultural losses, e.g., those caused by frosts, floods	Cooperates with central government and local authorities; farmers and producers organizations, rural NGOs, and communities

(continued)

Table 3.3 (continued)

Institution, organization, network	Type	Focus areas (from the point of view of MAPs producers)	Connectivity and status
Herbal companies	Industry	Contracting MAPs plantations with farmers; advisory services for farmers concerning MAPs cultivation and postharvest treatment; advisory services for owners of local collection points, purchasing wild-growing plants; production of MAPs seeds and/or seedlings; MAPs postharvest treatment (drying, cleaning and packaging); control and evaluation of herbal raw material and seeds quality before sale; MAPs storage; herbal raw material processing and manufacturing; import and export of MAPs raw materials and products thereof	Cooperates with farmers, collection points, other MAPs companies, analytical laboratories, local government organizations, SPHPSIS, ARMA, agricultural advisory Centres, EUROPAM, PKZ, and producers of food, cosmetics, medicines, or agricultural products
Collection points	Local industry	Purchasing herbal raw materials obtained from wild-growing plants directly from local collectors; postharvest treatment of herbal raw materials, especially drying; training of herb collectors in terms of the collection area, MAPs identification, and collection methods (especially regarding herbal raw materials intended for organic production)	Herbal companies, herb collectors, certification bodies, and state sanitary inspection

(continued)

Table 3.3 (continued)

Institution, organization, network	Type	Focus areas (from the point of view of MAPs producers)	Connectivity and status
Commercial analytical laboratories	Private companies	Services of analysis of herbal raw materials and the products thereof on the content of pesticide residues, heavy metals, pyrrolizidine and terpene alkaloids, mycotoxins, microbiological contaminants, and other	Services for farmers, MAPs companies, certification bodies, inspection of the commercial quality of agricultural and food products, and state sanitary inspection
Agriculture universities and institutes	Academic and research organizations	Conducting research in agricultural practice–solving problems related to herbal production on demand of MAPs companies or the Ministry of Agriculture and Rural Development; development of norms, standards and consultations for the needs of state inspections and the Ministry of Agriculture and Rural Development; caring out workshops and conferences for the ministry, certification bodies, MAPs companies, farmers and herb collectors; providing knowledge and plant materials for cultivation; caring out national and international projects	Cooperation with farmers, MAPs companies, certification bodies, Ministry of Agriculture and Rural Development, inspection of the commercial quality of agricultural and food products, agricultural advisory Centres, EUROPAM, PKZ, ECPGR
Farmers	Private sector—Production	Exchange of knowledge and means of production; herb drying services; rental of machinery and tools for cultivation and harvesting	Other farmers
The polish herbal committee	Professional association	Advocacy, legislation, and industry networking on the country level	Network node

(continued)

Table 3.3 (continued)

Institution, organization, network	Type	Focus areas (from the point of view of MAPs producers)	Connectivity and status
The European herb growers association	Professional association	Advocacy, legislation, and industry networking on the European level	Network node
MAPs working group–ECPGR (the European cooperative Programme for plant genetic resources)	Collaborative programme among European countries	Ensuring the long-term conservation and facilitating the increased utilization of plant genetic resources	Agriculture, universities and institutes, farmers, and herbal companies

2025). More information on the cooperation of the Regional Agricultural Adivisory Centres and MAPs producers is presented in Table 3.3.

4. National research institutes and agricultural universities involved in MAPs production

Institute of Natural Fibres and Medicinal Plants (Instytut Włókien Naturalnych i Roślin Zielarskich, IWNiRZ)—National Research Institute is the governmental research Institute working on fibre species and medicinal and aromatic plants.

Main directions of activity regarding MAPs are:

- breeding of varieties of fiber and herbal plants, their agrotechnology and preliminary processing;
- recipes and technologies of Polish herbal medicinal products;
- pharmacological and microbiological analysis of herbal raw materials, preparations, and technical materials;
- comprehensive analysis of biologically active substances in terms of quality assessment;
- expertise, consultations, and industry advice for state and local government bodies and business entities;
- cooperation with Polish GeneBank concerning MAPs;

The Institute is also Poland's official contact point for European platforms like the ECPGR–MAPs Working Group.

Warsaw University of Life Sciences (WULS—SGGW) is the oldest agricultural university in Poland. The school is a thriving academic center carrying out research and collaborating with both governmental institutions and the industry. The studies on the MAPs production and their processing are carried out here mainly in the Department of Vegetable and Medicinal Plants. This unit cooperates with the Ministry of Agriculture and Rural Development, Certification Bodies, and the Inspection of the Commercial Quality of Agricultural and Food Products. The Department also works with herbal companies concerning organic production of MAPs originating from cultivation and wild-growing species, preparing norms, standards, and workshops for all the stakeholders involved in this type of production. Diverse scientific projects are carried out in the Department, ranging from introducing wild-growing, endangered MAPs into cultivation, studies on biological progress in MAPs production, their breeding, and research on the quality of herbal raw materials and their biological activity. In this respect, the Department collaborates with the Medical University of Warsaw. The Department also cooperates with:

- farmers producing MAPs,
- herb collectors,
- herbal companies carrying out the commissioned research and expertise,
- Central and Regional Agricultural Advisory Centres (workshops for advisors, implementation projects),
- ECPGR (a member of MAPs Working Group),
- Polish GeneBank—since the 1990s, the Department has been carrying out the collection of MAPs within the Polish GeneBank,
- foreign research institutions, such as the University of Bangalore, the Natural Research Centre in Cairo, or the University of Life Sciences in Ulaanbaatar.

5. Industry-oriented organizations

Polish Herbal Committee (Polski Komitet Zielarski) (Polish 2025), which is the main organization representing the interests of the Polish herbal industry since 1930. The Committee includes representatives from pharmaceutical companies, distributors, scientific institutes, herbal plant growers, and individual practitioners. PKZ has over 200 ordinary members and 20+ supporting enterprises. The main activities are shaping public opinion, influencing legislation on herbal products, collaborating with European and international bodies, and supporting the development and standardization of herbal practice and medicine in Poland. The Committee is recognized as the professional association and advocacy body for MAPs in Poland.

EUROPAM (the European Herb Growers Association) is the largest organization representing European herb producers, with Polish representatives (farmers and herbal companies' representatives) as members of this Association. EUROPAM represents the interests of its members regarding wild collection, growing, first processing steps, and quality assurance of MAPs. The Association provides guidance documents and databases to help its members better fulfill legal and GACP criteria.

ECPGR (The European Cooperative Programme for Plant Genetic Resources) aims at ensuring the long-term conservation and facilitating the increased utilization of plant genetic resources in Europe. Polish experts are members of the ECPGR Medicinal and Aromatic Plants Working Group, which contributes to the development of the conservation strategy of MAPs at the European level. The Group deals with legislative limitations implemented by EU trade regulations on endangered MAPs and with low knowledge of biological conditions and biodiversity status. The MAPS Working Group consists of European experts and scientists (including Polish experts from IWNRZ and WULS-SGGW).

3.7.1.3 MAPs Value Chains in Poland

Existing MAPs value chain representatives in Poland are farm input suppliers, industries, the machinery sector, farmers, herb collectors, postproduction services, processors, traders, consumers, policy makers, and higher education.

1. Farm input suppliers & industries

- agrochemical and fertilizer suppliers—major Polish players such as Grupa Azoty providing fertilizers;
- seed producers, including the Institute of Natural Fibers and Medicinal Plants—supplying improved herbal plant seeds to growers;
- plant protection products—while the MAPs value chain emphasizes non-chemical weed control, suppliers do not provide permissible organic amendments and biological solutions.

In Poland, plant protection products and fertilizers used in the cultivation of MAPs are standard agricultural products. There are no special pesticides or fertilizers intended specifically for this sector. Due to the withdrawal of many plant protection products from use, a significant problem has arisen for herbal production. It is very difficult to control weeds in large herb plantations without chemical protection (manual weeding costs are very high). Mechanical treatments alone are not sufficient in the long term. Further withdrawals of plant protection products may lead farmers to abandon this particularly demanding type of production. This issue is also linked to the very strict quality assessment of herbal raw materials. These raw materials are evaluated, among other, for pesticide residues, microbiological and mineral contamination, the presence of pyrrolizidine and tropane alkaloids, mycotoxins, and heavy metals. Meeting such standards is sometimes impossible, which reduces the profitability of this production and occasionally leads to financial losses.

In Poland, farmers mostly produce their own seed material of MAPs, or sometimes import seeds from abroad. To some extent, seeds or seedlings are available from herbal companies (e.g., FZL Sp. z o.o., Martin Bauer Poland), but the range is limited. Large seed companies are not interested in breeding or producing seed material of MAPs, as this type of production is globally marginal (many species are cultivated on relatively small areas) and very cost-intensive.

2. Machinery sector

   - agricultural machinery manufacturers—firms like PRONAR, SAMASZ, and others manufacture equipment for planting, harvesting, drying, and primary processing, tailored to niche and specialty crops like MAPs;
   - precision agriculture tools—suppliers of smart agriculture solutions (including Polish tech start-ups and agricultural service providers) are increasingly providing precision tools with the capacity to be used for MAP cultivation;
   - suppliers of machines for cutting, grinding, granulating, sorting, and cleaning herbal raw materials, both dry and fresh (example of SCORPION firm);

- producers of elements of drying systems (example of BIN Sp z o.o.).

3. Farmers & collectors

   - over 10% of Polish farms grow MAPs, with both large-scale and smallholders engaged, especially in southeastern and central regions;
   - collectors of wild-growing MAPs still play an important role, but there is a shift toward cultivation for quality and sustainability;
   - "farmers marketing groups" and small grower partnerships facilitate group marketing and resource sharing;
   - postproduction services;
   - drying and sorting facilities—many are farm-integrated or managed by local herb partnerships; larger processors may operate centralized facilities.

4. Processors

   - Herbapol companies are the national leaders in processing MAPs into teas, extracts, supplements, and plant-based pharmaceuticals;
   - smaller companies and start-ups are also active in the development of functional foods, cosmetics, and essential oils.

5. Traders & export/import companies

   - herbal raw material companies—purchasing dried raw material from farmers and wild-growing plants from collectors, then distribute to processors, pharmacies, or for export; many operate as intermediaries for large pharmaceutical and food companies across Europe;
   - Polish MAPs exporters are among the EU's one of the largest, serving especially Germany, UK, France, Italy, and the USA.

6. Consumers

   - Polish consumers are highly quality-and eco-conscious; there is a rising demand for organic, sustainably sourced herbs and products;
   - a large segment of consumption are supplements, teas, and natural health remedies, with online sales growing rapidly.

7. Policy makers & regulators

   - Ministry of Agriculture and Rural Development, which oversees agricultural practices, subsidy schemes, and MAPs registration;
   - Polish Herbal Committee (Polski Komitet Zielarski, PKZ), which is the main industry advocacy group, also shapes regulations and sector standards;
   - Polish research institutes like IWNiRZ work with European platforms such as ECPGR for conservation and utilization policy.

8. Higher education & research

   - Universities—Warsaw University of Life Sciences, Poznań University of Life Sciences, among others, conduct research, breeding, and offer MAP-focused education;

- IWNiRZ—national research backbone for MAPs genetics, breeding, and cultivation;
- Fundacja Instytut Zielarstwa Polskiego i Terapii Naturalnych—Foundation Institute of Polish Herbalism and Natural Therapies.

These actors from the 8 groups collectively ensure Poland's vital status in the European MAPs sector, integrating production, quality, research, and export within an advanced and well-regulated value chain.

3.7.2 *Türkyie*

3.7.2.1 Turkish MAPs System

Türkiye produces many plant-based products used as inputs in the pharmaceutical, plant chemicals, food and food additives, cosmetics, and perfumery sectors. These products, which are collected from nature and cultivated, must be available in the desired quantity and sufficient quality to adequately address the market potential of medicinal and aromatic plants. Developing crop varieties, determining suitable ecological conditions, harvesting natural plants on time without harming nature, and determining post-harvest processes and technologies—all are important factors in creating quality products that meet consumer and industry demands and in developing production and market opportunities for medicinal and aromatic plants.

Many MAPs species are known in the country, mostly from the flora, but also including other cultivated plants. There were in 2015 about 21.678 thousand ha used for collection in Türkiye. Generally, they are public.

3.7.2.2 Medicinal and Aromatic Plants Networks in Türkiye

Türkiye has a network of MAPs consisting of cooperatives, associations, ministries, research institutes, and industry-focused organizations.

The most important entities creating national networks and structures are as follows:

1. The Ministry of Food, Agriculture and Livestock and the Ministry of Health, which are controlling and licensing the herbal medicine sector in Türkiye.
2. Institute of Medicinal and Aromatic Plants—National Research Institute established in Antalya (laboratory, collection garden-production, propagation greenhouse) and in operation since 2010. The main purpose of these Institute is to develop the registered varieties by cultivating new plant species for the production of standard and high quality raw materials required by different industries in medicinal, aromatic and dye plants in Türkiye. The branches of the Institute which are located in different regions of Türkiye, deal with selecting the appropriate type of culture to their ecology and engage in breeding and variety

development work. Improvement studies have been made in order to effectively integrate quality studies to determine the active substances in plants. As a result of these studies, 41 cultivars of medicinal and aromatic plants belong to 15 species have been registered till now. The production areas of these species are increasing every year. In the following process, it is aimed to continue the culture, breeding, cultivation and quality studies and to develop new varieties and increase the production and usage areas of these varieties. On the other hand, in order to get more shares from the world herbal medicine market which is expected to reach 5 trillion USD in 2050, it is planned to give priority to the export of processed products instead of raw material export.

3. Herbal Companies

According to the data obtained for 2025–1470 companies were found to be commercially engaged in export-import and domestic operations in Türkiye. Some are wholesalers/distributors, others are retailers and raw material suppliers. They are active in particular in sectors such as food, cosmetics, herbalist and medicinal sector.

4. National and Regional Research Networks

Some institutions provide training and assessment services on topics such as collecting, production, harvesting, and post-harvesting of MAPs, and some municipalities also offer support for the expansion of production:

- Agricultural Research Institute provides information and training;
- Provincial Directorate of Food, Agriculture, and Livestock provides information, education and nurseries activities;
- General Directorate of Forestry provides information and education;
- Municipalities provide services such as seedlings-nurseries-distillation and extraction-equipment-information-education.

Technical information is disseminated to producers in different formats: promotional activities, supported by projects (training will be included in the project), workshops, brochures, seminars, field days, flora tours, books available on websites.

3.7.2.3 MAPs Value Chains in Türkiye

The entities engaged in MAPs wild collection are farmers, especially women farmers, farmer organizations, industrialists, traders who export or import MAPs products and the cooperatives. The wild MAPs are transformed with drying and distillation. The majority of collectors do the bulk trade, but a part of collectors can sell retail.

As for consumers—demand among consumers in Türkiye for plants and products sourced from organic, sustainable sources is increasing. This is mostly thanks to the growing awareness of the topic. Consumers are increasingly choosing supplements, teas, and natural health products. In addition, the number of herbal prescription drugs sold in pharmacies has increased significantly and demand for them is comparable to other drugs.

The use of spices is also very widespread in the country. The MAPs plantations have increased every year in Türkiye. The property of the plantations could be private or rented, belonging to a cooperative or sometimes constituting contracted farming. But the maintenance of the crops is usually family business. The type of production used could be conventional or organic, following good agricultural practices. The main raw material obtained from the MAPs cultivation, is fresh and dry herb.

The MAPs plantations have been established mostly in the last 10 years. More than 98% are privately owned. Some of the producers use conventional techniques, while other use organic techniques.

3.7.3 Bulgaria

3.7.3.1 Bulgarian MAPs System

Bulgaria has a long-standing tradition in the cultivation and processing of medicinal and aromatic plants.. Due to its specific geographical location, Bulgaria exhibits considerable heterogeneity in relief, geological structure, and soil composition. The distinct microclimatic conditions contribute to high levels of biodiversity and determine the specific phytochemical and species composition of medicinal plants. This unique profile renders Bulgarian herbal raw materials highly valued on the global market.

The structure of the Bulgarian MAPs system is presented in Table 3.4. The sector operates through various legal forms: sole trader (ET), Single-Member Limited Liability Companies (Ltd. = EOOD/OOD in Bulgaria), cooperatives, and non-profit associations and is financed mainly through own resources, shareholders, and project support. The scale of operations depends on utilized agricultural area (UAA), labour input, farm size, and market orientation (Bondev 1995).

The establishment of regulatory frameworks and the oversight of farmers in relation to state funding are carried out by the Ministry of Agriculture and Food (MAF), and of Ministry of Environment and Water (MOEW), while the quality control of raw materials is ensured by the Bulgarian Food Safety Agency (BSFA). Scientific support and specialized training in the field of MAPs are provided by the research institutes of the Agricultural Academy, Universities (e.g.the Agricultural University in Plovdiv; University of Food Technology), and the Bulgarian Academy of Sciences (Section of Applied Botany, and Institute of Organic Chemistry with the Centre of Phytochemistry). In Bulgaria, there are several professional organizations and associations that represent, to varying degrees, the interests of herbal collectors and farmers engaged in the cultivation of medicinal and aromatic plants. However, the sector remains relatively fragmented, as it includes a large number of individual collectors and small-scale producers, what obviously limits the level of centralized representation.

The Bulgarian Association of Herbalists and Mushroom Collectors (https://babg.eu/) is among the most directly relevant organizations, bringing together collectors,

Table 3.4 Bulgarian MAPs system

Organization	Type	Main responsibilities
Ministry of environment and water (MOEW)	State authority	– Implements medicinal plants act-sets annual harvesting quotas-issues permits for wild collection-protects endangered species; it regulates the activities of processing of herbs;
Ministry of Agriculture and food (MAF)	State authority	– Oversees cultivation of medicinal plants-develops agricultural policies-supports farmers through programs/ subsidies
State fund agriculture	State agency	– Provides financial support and subsidies-manages rural development programs
Regional inspectorates of environment and water (RIEWs);	Regional state bodies	– Monitor harvesting activities-issue/verify permits locally-conduct inspections and enforce regulations
Bulgarian food safety agency (BFSA)	Control authority	– Ensures safety and quality of herbal products-inspects processing facilities-controls food supplements and herbal teas
Executive Agency for Medicines (BDA / IAL)	Regulatory authority	– Authorizes herbal medicinal products-controls pharmaceutical use of plant-based medicines
Bulgarian Association of Herbal Producers and Exporters (BAHPEE)	Branch organization	– Represents producers, collectors, exporters-Supports market access and exports-Promotes industry standards
National Branch Organization for essential oils, perfumery and cosmetics (NBIOPC)	Branch organization	– Represents producers of essential oil crops (rose, lavender, etc.)-supports industry development and policy dialogue
Bulgarian chamber of commerce and industry (BCCI)	Business organization	– Facilitates trade and export partnerships-provides business support services
Local cooperatives and private companies	Producers	– Collect, cultivate, process medicinal plants-supply domestic and export markets
EU/international frameworks (GACP, Natura 2000, CITES)	Supranational	– Set standards for sustainable harvesting-protect biodiversity and endangered species-regulate international trade
Analitycal laboratories	Private companies; mixed (public infrastructure, commercial model); public (state-funded)	Services for farmers, MAPs businesses, certification, and quality & sanitary inspection of agricultural and food products

(continued)

Table 3.4 (continued)

Organization	Type	Main responsibilities
Agriculture university and Institute of Roses, essential and medical cultures; Institute of Biodiversity and Ecosystem Research at the Bulgarian Academy of Sciences	Academic and research organizations, state institutions	Conduct scientific research across all aspects of medicinal plants; provide consultations, advisory services, and scientific support to companies and herb producers, including upon request from the Ministry of Agriculture and food; deliver guidance on the collection and cultivation of medicinal plants; develop standards and provide expert evaluations; participate in committees for establishing the resource of wild herbs, as well as in developing strategies for the protection of rare and endangered herbs. Design and optimize cultivation technologies; and implement national and international projects
Farmers, herb collectors and herbalist	Private sector—Producers, herbs collectors and herbalist	Knowledge exchange; herb drying and processing services
Bulgarian association of herbalists and mushrooms	Profesional association	Voluntary, independent association uniting individuals and legal entities working in the field of herb and mushroom production, as well as in the trade in herbs, mushrooms and wild fruits and assisting in the protection of their professional and material interests

processors, and traders of MAPs and participating in discussions on the regulatory framework, including the Law on Medicinal Plants. Another key organization is the Bulgarian National Association of Essential Oils, Perfumery and Cosmetics (www.bnaeopc.com), which represents producers of essential oil crops such as lavender, rose, and lemon balm, encompassing both farmers and processors. In addition, there are local cooperatives and community-based associations, particularly in regions such as the Rhodopes and the Balkan Mountains, that play an important role in organizing collection, processing, and marketing activities at the local level.

3.7.3.2 Medicinal and Aromatic Plants Networks in Bulgaria

The activities within the Bulgarian medicinal and aromatic plants (MAPs) sector can be grouped as follows: (1) farmers that cultivate MAPs and sell the raw material; (2) integrated models combining cultivation, processing (EO, floral waters, concretes, absolutes, and dry extracts), and export; (3) harvesting of wild medicinal plants and their subsequent processing by herbalists; (4) Agricultural Advisory Centres and State Supervisory Bodies; (5) Research institutes and agricultural universities, (6) industry-oriented organizations.

1. Farmers that cultivate MAPs

Bulgarian farmers involved in MAPs cultivation are typically small–to medium-scale agricultural producers who cultivate MAPs primarily for sale as raw material rather than finished products. The spatial distribution of these farmers is uneven, with higher concentrations in mountainous and semi-mountainous regions such as the Rhodope and Balkan Mountains, as well as in specialized agricultural zones like the Rose Valley, Tracian lowland ect. In Bulgaria, alongside the collection of wild medicinal plants, a substantial proportion of species are also widely cultivated. Wild harvesting is typically limited to species for which established cultivation technologies are lacking or that are of limited commercial interest, or there is a high abundance of them. In contrast, species of significance to the pharmaceutical, cosmetic, and food and flavor industries are traditionally subject to systematic cultivation. Traditionally fermers, produce: Fresh or dried herbs; Raw plant parts (flowers, leaves, roots, seeds). Their main buyers are: Processors (distilleries, extract producers); Export companies, Pharmaceutical and cosmetic industries, and shops. Agricultural holdings cultivating MAPs in Bulgaria varied in size depending on the type of crop. In the case of traditional and labor-intensive crops, such as *R. damascena* EO and various herbs, farms are typically small to medium-sized (0.05–10 ha). In contrast, more highly mechanized crops, such as *L. angustifolia* and *Ch. recutita*, are associated with larger holdings, often exceeding 50 ha. This structural pattern reflects both the biological characteristics of the crops and the degree of mechanization, as well as the market orientation of production.

2. Integrated models combining cultivation, processing (EO, floral waters, concretes, absolutes, and dry extracts), and export

The combination of growing medicinal plants and processing is a tradition in Bulgaria known from the past. Some of the earliest records of integrated models of Bulgarian *R. damascena* EO cultivation and exports to Marseille, London, and New York date back to 1720. The first private rose distillery for the processing and trade of *R. damascena* EOs was established in 1820. Professional processing methods and systematic quality analysis of rose EO began to develop in 1907. The cultivation and processing of MAPs in Bulgaria have a long-standing tradition, which has provided the foundation for the development of modern agricultural and processing enterprises. During the socialist period (1945–1989), the collection and processing of medicinal plants were centrally organized and carried out by state enterprises and cooperatives. After the pass of this period, privatization allowed private businesses to emerge and many herb-processing and trading companies were founded (Ltd.). Most companies were founded after 1990 as family-owned farms (sole trader). Generally, the sector combines centuries-old practices with modern industrial processing and export-oriented business models. The number of workers per enterprise varies widely, from 5 to 150, depending on the production scale. The major Bulgarian companies are presented in Table 3.5.

3. Harvesting of wild medicinal plants and their subsequent processing by herbalists

Overall, Bulgaria's flora comprises approximately 4000 plant species, of which around 700–780 are recognized as medicinal (Evstatieva et al. 2007). Between 200

Table 3.5 Major Bulgarian herbal companies

Company	Main activity	Specialization	Website
Galen-N ltd.	Production, manufacturers, export	Producers and manufacturers of Bulgarian rose, lavender, chamomile, lemon balm (Melissa) and other essential oils, floral waters, concretes and absolutes.	https://galen-n.bg
Botanical EU ltd.	Processing & export	Dried herbs, teas	https://botanical.bg/en/pages/about-us?utm_source=chatgpt.com
Alta oils ltd	Production, manufacturers, export	Producers and manufacturers of Bulgarian rose, lavender, chamomile, lemon balm (Melissa) and other essential oils, floral waters	https://www.altaoils.com/
Vesselino ltd.	Export & trade	Medicinal herbs, essential oils	https://vesselino.com/?utm_source=chatgpt.com
Mountain rose Bulgaria ltd.	Production & export	Organic herbs, extracts, oils	https://mountainrosebulgaria.com/?utm_source=chatgpt.com
Biofresh ltd.	Cosmetics production	Rose oil & herbal cosmetics	https://biofresh.bg/en/about/?utm_source=chatgpt.com

and 300 medicinal plant species are collected from wild populations. Wild collection remains an important first step in the value chain, as many Bulgarian MAPs species are still harvested from natural habitats across national parks and economically important mountainous regions. The country has a well-established tradition of exporting herbal raw materials, and MAPs continue to be positioned strongly on international markets due to their purity and high active ingredient content. Increasingly, processing companies and exporters emphasize sustainability and quality standards, ensuring that herbs retain their potency and authenticity throughout the supply chain. In Bulgaria, the Ministry of Environment and Water (MOEW) annually determines quotas for the collection of medicinal plants, while oversight of the implementation of these regulations is delegated to the Regional Environmental Inspectorates (Bulgarian 2026). Furthermore, it regulates the activities of herbal collection and processing entities in order to ensure sustainable use and conservation of plant resources.

According to the requirements of MOEW (Act 2026) before herb collection, permits must be obtained, and state fees must be paid. The harvesting of wild herbs is most often carried out by less qualified individuals, who typically receive instructions from herbalists. In addition, the conditions that an herbalist must meet, along with the documentation that must be completed, are often difficult to fulfill for collectors, who are usually people from large families or pensioners.

4. Agricultural Advisory Centres and State Supervisory Bodies in Bulgaria

Agricultural advisory and regulatory oversight of medicinal plants in Bulgaria are embedded within a comprehensive institutional framework that integrates agricultural policy, environmental protection, and public health considerations. Advisory services are primarily provided by the National Agricultural Advisory Service (NAAS) (Implementation 2026), which operates under the Ministry of Agriculture and functions as the central component of the national Farm Advisory System. Through its regional network, NAAS supports farmers, including those engaged in the cultivation of medicinal and aromatic plants, by offering technical guidance, promoting sustainable practices, and facilitating compliance with European Union standards. In parallel, regulatory and supervisory responsibilities are distributed among several state bodies. The Ministry of Agriculture and Food oversees agricultural production and plant health, while the Bulgarian Food Safety Agency, as its successor structure in phytosanitary control, ensures compliance with plant protection and safety requirements. The Ministry of Environment and Water regulates the sustainable use of wild medicinal plant resources by establishing harvesting quotas, maintaining lists of protected species, and enforcing biodiversity legislation. Additionally, the Ministry of Health supervises the use of medicinal plants in pharmaceutical and food products, ensuring adherence to safety and quality standards. This multi-level governance system, supported by key legislative instruments such as the Medicinal Plants Act and the Biological Diversity Act, reflects a coordinated approach that balances economic utilization with the conservation and sustainable management of medicinal plant resources.

5. National Research Institutes and Agricultural Universities in Bulgaria involved in MAPs production

Research and higher education activities related specifically to medicinal plants in Bulgaria are supported by a well-established network of national research institutes and universities, which contribute to the scientific basis and sustainable development of the sector. A leading role is played by the Agricultural Academy, which coordinates applied research through specialized institutes engaged in plant breeding, genetic resources, and cultivation technologies relevant to medicinal species. Within this framework, the AgroBioInstitute conducts studies on plant biotechnology, germplasm conservation, and the enhancement of biologically active compounds in medicinal plants. A particularly important contribution to the study of essential oil-bearing medicinal species is made by the Research Institute for Roses, Essential and Medicinal Plants in Kazanlak, which specializes in breeding, cultivation technologies, and the chemical characterization of aromatic species. Complementing these efforts, the Bulgarian Academy of Sciences hosts several key research units. Notably, the Institute of Organic Chemistry with the Centre of Phytochemistry is internationally recognized for its work on the isolation and characterization of bioactive compounds derived from medicinal plants. In addition, the Institute of Biodiversity and Ecosystem Research conducts comprehensive scientific studies encompassing all aspects of medicinal plant diversity, ecology, and sustainable use. Additional contributions are provided by the University of Food Technologies, where research focuses on the processing, quality, and application of plant-derived products, including essential oils, in the food and related industries. Higher education institutions, particularly the Agricultural University—Plovdiv, as well as the University of Forestry and Sofia University "St. Kliment Ohridski," further support the sector through specialized education and research in phytochemistry, pharmacognosy, plant physiology, and the cultivation of medicinal and essential oil crops. Through the integration of fundamental and applied research with academic training, these institutions play a critical role in advancing knowledge, improving cultivation and processing techniques, and ensuring the sustainable use and valorization of medicinal plants and their essential oil derivatives in Bulgaria.

6. Industry-oriented organizations

In Bulgaria, industry-oriented organizations in the field of MAPs are primarily represented by sectoral associations, integrated production enterprises, and cooperative structures that link cultivation, processing, and export activities. The most prominent national body is the Bulgarian National Association of Essential Oils, Perfumery and Cosmetics (BNAEOPC), which functions as a key industry cluster representing producers, processors, and exporters of essential oils and aromatic crops. The association also facilitates cooperation with research institutions and maintains international links with European and global organizations in the sector, thereby contributing to standardization and market integration.

Alongside this representative organization, the sector is characterized by a large number of privately owned and vertically integrated companies engaged in the cultivation, distillation, and export of MAPs and other essential oil-bearing species. These enterprises are often organized into regional production networks or cooperatives, which play an important role in raw material aggregation and primary processing. Overall, the Bulgarian MAP industry is structured around a combination of strong industry associations, export-oriented enterprises, and cooperative production systems, supported by collaboration with scientific and research institutions. A key driver of modernization within the value chain is Bulgaria's Center of Competence for Sustainable Utilization of Bio-Resources from Medicinal and Aromatic Plants (Center 2026) which integrates research, green extraction technologies, and product development to support pharmaceutical, nutraceutical, and cosmetic industries. The Center operates along the entire bio-resource value chain from biomass characterization to final product formulation strengthening links between science and industry. Such initiatives enhance the competitiveness of Bulgarian MAPs products while promoting rural development and environmentally responsible harvesting practices. Additionally, legislation and long-term biodiversity studies guide sustainable collection and trade, ensuring that MAPs resources remain viable under growing commercial demand. Overall, Bulgaria's MAPs value chain blends traditional knowledge, strong biodiversity assets, and emerging innovation infrastructures, positioning the country as a regional leader in herbal production and bio-based product development.

3.7.3.3 MAPs Value Chains in Bulgaria

Medicinal plant value chains in Bulgaria are characterized by a well-developed and export-oriented structure, integrating both wild collection and cultivated production systems. The value chain begins with two primary sources of raw material: the harvesting of wild-growing medicinal plants and the cultivation of selected species. At the upstream level, the sector is supported by farm input suppliers and related industries, which provide essential resources such as certified seeds and planting material, fertilizers (both organic and mineral), plant protection products, irrigation systems, and specialized machinery for cultivation and harvesting. These inputs are often developed and supplied in collaboration with research institutes and breeding centers, ensuring the suitability of plant varieties and technologies for medicinal use. Farmers rely on these inputs alongside advisory services to implement good agricultural and collection practices, including soil management, crop rotation, pest control, and optimal harvesting techniques, all of which are critical for maintaining the quality, safety, and traceability of medicinal plant raw materials.

Following primary production, the value chain proceeds to initial processing stages such as cleaning, sorting, drying, and storage, frequently carried out by farmers, collectors, or small enterprises. For essential oil-bearing medicinal plants, including rose and lavender, this stage also involves distillation, representing a significant value-adding activity closely linked to specialized processing industries. Secondary processing and industrial transformation follow, where raw materials are converted into higher-value products such as dried herbs, extracts, tinctures, essential oils, and active pharmaceutical ingredients. These processes are undertaken by companies in the pharmaceutical, cosmetic, and food industries, which impose strict requirements for quality control and standardization.

The final stages of the value chain involve distribution and marketing, with Bulgaria maintaining a strong position as an exporter of medicinal plant raw materials and processed products to European and global markets. Despite its strengths, the sector faces challenges related to the fragmentation of primary producers, variability in input use and raw material quality, and limited value addition at the farm level. Nevertheless, the integration of farmers, farm input suppliers, processing industries, research institutions, and regulatory bodies contributes to a relatively robust and adaptable value chain, supporting the sustainable development and international competitiveness of the medicinal plant sector in Bulgaria.

Generally, there is no specialized equipment for medicinal plants. The equipment used in other branches of plant growing is usually used. Machinery used for seedling and propagation of medicinal plants mainly includes seed sowing and tray filling machines that ensure uniform and efficient planting of seeds in nursery trays. Propagation systems in greenhouses, such as misting units, heating mats, and grow lights, help maintain ideal conditions for germination and early plant growth. For large-scale production, tissue culture equipment is used to produce disease-free and genetically identical medicinal plants. Transplanting machines then move seedlings from trays to fields or pots quickly and with minimal damage. Overall, these technologies improve efficiency, survival rates, and consistency in medicinal plant production. In Bulgaria, equipment comes through distributors such as Interagri, Bulagro, and Agrikol, which supply major European and global brands such as New Holland, Pöttinger, Kinze, Netafim, and Dutch greenhouse automation systems.

In Bulgaria, MAPs cultivation uses fungicides, insecticides, and herbicides for plant protection, along with mineral fertilizers (NPK and micronutrients) in conventional systems. Overall, there are no plant protection products specifically registered for use on MAPs within conventional agricultural systems. Furthermore, specific weed control technologies have not been established in organic agriculture. However, there is a strong shift toward organic farming, where growers use bio-pesticides, compost, and biostimulants instead of chemicals.

In organic farming, weed management prioritizes mechanical and physical methods such as hand weeding, inter-row cultivation, flame or steam weeding, mulching, and thermal treatments. Likewise, they use biological approaches such as living mulches or the application of spent biomass as organic mulch. Farmers also adopt advanced practices, including crop rotation, cover cropping, mowing, soil solarisation, mechanical harvesters equipped with weeding units, and precision agriculture tools such as drone-based weed mapping and sensor-guided cultivators.

3.8 Discussion and Conclusions

Stakeholder engagement constitutes a foundational determinant of coherence, sustainability, and competitiveness within the medicinal and aromatic plants (MAPs) sector. Given the sector's inherent complexity—encompassing cultivation, wild collection, processing, certification, research, and governance—its effective functioning depends on coordinated contributions from a heterogeneous set of actors. Structured and systematic engagement mechanisms help mitigate value-chain fragmentation by fostering transparency, improving information flows, and aligning the diverse interests, capacities, and expectations of stakeholders. Such engagement strengthens the integration of scientific knowledge into practice, enhances the uptake of good agricultural and collection practices, and supports the transition toward more sustainable production systems, including non-chemical weed management. Moreover, collaboration between private actors and public institutions ensures that regulatory frameworks and support schemes are responsive to sectoral needs, thereby enhancing policy effectiveness and enabling better access to funding, data, and innovation infrastructures. Academic and research organizations play a crucial role by advancing evidence-based practices, building technical capacity, and supporting the development of new technologies and value-added applications. Inclusive engagement also reinforces social equity by amplifying the participation of smallholders, collectors, women, and emerging producers whose contributions are essential for resilient rural economies. In the context of growing international demand and increasing sustainability requirements, robust and well-coordinated stakeholder engagement is not merely advantageous but essential for ensuring the long-term viability of MAPs resources and the continued evolution of competitive, innovation-driven value chains. Ultimately, the sector's future trajectory will depend on its ability to institutionalize collaborative processes that promote shared learning, collective problem-solving, and mutual accountability.

References

Act (2026). Act on medicinal plants and the ordinance on the collection and drying of medicinal herbs, 2026. Accessed 14 May 2026. https://www.moew.government.bg/en/nature/legislations/national-legislation/laws/

Agriculture (2025). Agriculture advisory Centre in Brwinów, 2025. Accessed 12 Sep 2025. https://en.cdr.gov.pl/index.php/about-us/our-mission

Amapseec (2025). Amapseec Association for Medicinal and Aromatic Plants of southeast European countries, 2025. Accessed 09 Sep 2025. http://amapseec.com/

Bondev I (1995) Flora and Medicine. Chorological atlas of medicinal plants in Bulgaria. Academic Publishing House "Prof. Marin Drinov", Sofia, Bulgaria, pp 9–10

Bulgarian (2026). Bulgarian Ministry of Environment and water, 2026. Accessed 14 May 2026. https://www.moew.government.bg/en/the-collection-of-herbs-for-commercial-purposes-from-24-species-of-medicinal-plants-from-their-natural-habitats-is-prohibited/

Center (2026). Center of Competence: sustainable utilization of bio-resources from medicinal and aromatic plants, 2026. Accessed 15 May 2026. https://lcpe.uni-sofia.bg/center-of-competence-bioresources/

Chandra P, Kumar J (2021) Linking the medicinal and aromatic plants business to sustainable resource management and economic prosperity: a value chain analysis. Area Dev Policy 6(4):470–482

EU CAP (2025). EU CAP network, 2025. Accessed 08 Sep 2025. https://ec.europa.eu/eip/agriculture/sites/default/files/fg35_mp1_mainactorsmarketscollaboration_2020_en.pdf

Europam (2025). Europam, the European herb growers association, 2025. Accessed 09 Sep 2025. https://www.europam.net/

European Cooperative (2025). European cooperative Programme for plant genetic resources, 2025. Accessed 08 Sep 2025. https://www.ecpgr.org/working-groups/medicinal-and-aromatic-plants

Evstatieva L, Hardalova R, Stoyanova K (2007) Medicinal plants in Bulgaria: diversity, legislation, conservation and trade. Phytologia Balcanica 13(3):415–427

Implementation (2026). Implementation of the EU farm advisory system in Bulgaria, 2026. Accessed 15Apr 2026. https://eu-cap-network.ec.europa.eu/projects/practice-abstracts/implementation-eu-farm-advisory-system-bulgaria_en#

International Council (2025). International Council for Medicinal and Aromatic Plants. Accessed 09 Sep 2025. http://icmap.org/?i=1

IUCN SSC (2025). IUCN SSC medicinal plant specialist group, 2025. Accessed 08 Sep 2025. https://iucn.org/our-union/commissions/group/iucn-ssc-medicinal-plant-specialist-group

Society (2025). Society for Medicinal Plant and Natural Product Research, 2025. Accessed 09 Sep 2025. https://ga-online.org/

Sucholas J, Ukhanova M, Greinwald A (2021) Wild collection of medicinal and aromatic plants (MAPs) in Poland for commercial purposes: a system reviewed. Herba Polonica 67(3)

Polish (2025). Polish herbal committee, 2025. Accessed 09 Sep 2025. https://www.pkz.pl/en/home-page/

Index

V. J. P. D. Martinho et al. (eds.), *Chemical-free Weed Control*,
SpringerBriefs in Agriculture, https://doi.org/10.1007/978-3-032-34699-5

GPSR Compliance
The European Union's (EU) General Product Safety Regulation (GPSR) is a set of rules that requires consumer products to be safe and our obligations to ensure this.

If you have any concerns about our products, you can contact us on

ProductSafety@springernature.com

In case Publisher is established outside the EU, the EU authorized representative is:

Springer Nature Customer Service Center GmbH
Europaplatz 3
69115 Heidelberg, Germany

www.ingramcontent.com/pod-product-compliance
Ingram Content Group UK Ltd.
Pitfield, Milton Keynes, MK11 3LW, UK
UKHW021639180926
475650UK00001B/40
* 9 7 8 3 0 3 2 3 4 6 9 8 8 *